SpringerBriefs in Mathematics

Series Editors
Nicola Bellomo, Torino, Italy
Michele Benzi, Pisa, Italy
Palle Jorgensen, Iowa City, USA
Roderick Melnik, Waterloo, Canada
Otmar Scherzer, Linz, Austria
Benjamin Steinberg, New York, NY, USA
Lothar Reichel, Kent, USA
Yuri Tschinkel, New York, NY, USA
George Yin, Detroit, USA
Ping Zhang, Kalamazoo, MI, USA

SpringerBriefs present concise summaries of cutting-edge research and practical applications across a wide spectrum of fields. Featuring compact volumes of 50 to 125 pages, the series covers a range of content from professional to academic. Briefs are characterized by fast, global electronic dissemination, standard publishing contracts, standardized manuscript preparation and formatting guidelines, and expedited production schedules.

Typical topics might include:

- A timely report of state-of-the art techniques
- A bridge between new research results, as published in journal articles, and a contextual literature review
- A snapshot of a hot or emerging topic
- An in-depth case study
- A presentation of core concepts that students must understand in order to make independent contributions.

SpringerBriefs in Mathematics showcases expositions in all areas of mathematics and applied mathematics. Manuscripts presenting new results or a single new result in a classical field, new field, or an emerging topic, applications, or bridges between new results and already published works, are encouraged. The series is intended for mathematicians and applied mathematicians. All works are peer-reviewed to meet the highest standards of scientific literature.

Titles from this series are indexed by Scopus, Web of Science, Mathematical Reviews, and zbMATH.

Vincenzo Basco

Nonsmooth Constrained Optimal Control on Infinite Horizon

A Brief Course

Springer

Vincenzo Basco
Acea Spa
Rome, Italy

ISSN 2191-8198 ISSN 2191-8201 (electronic)
SpringerBriefs in Mathematics
ISBN 978-3-032-08937-3 ISBN 978-3-032-08938-0 (eBook)
https://doi.org/10.1007/978-3-032-08938-0

This Springer imprint is published by the registered company Springer Nature Switzerland AG
The registered company address is: Gewerbestrasse 11, 6330 Cham, Switzerland

To Matteo Nicola and Andreea.

Preface

Infinite horizon control problems with constraints occupy a central position in the applied sciences, providing a rigorous mathematical framework for optimizing systems over unbounded time intervals. In the nonsmooth constrained setting, one must consider both state and control bounds together with the possible lack of differentiability of the dynamics or the cost functional. Such nonsmooth features—arising from switching rules, actuator saturation, or piecewise-defined penalties—require the use of tools from nonsmooth analysis, such as generalized gradients, subdifferentials, and viability theory. These ideas naturally interact with reinforcement learning, where agents face control limitations and nondifferentiable reward structures while aiming for long-term performance. In this way, infinite horizon control theory enriched by nonsmooth methods offers a framework for achieving asymptotic stability, optimality, and sustainable performance in complex dynamical systems.

The goal of optimal control theory is to establish necessary and sufficient conditions ensuring that a control minimizes or maximizes a given cost functional. Controls fulfilling these conditions are called optimal, and when represented in feedback form they yield an optimal synthesis. The construction of such controls relies on fundamental results such as Pontryagin's maximum principle and Bellman's dynamic programming principle. Pontryagin's principle, formulated in the 1950s, extends classical variational results like the Euler-Lagrange equation and the Weierstrass condition, asserting that the Hamiltonian must attain an extremum with respect to admissible controls. Bellman's approach, developed in parallel, complements it by expressing optimality through the Hamilton-Jacobi-Bellman (HJB) equation that characterizes the value function. Both frameworks have been extended to problems involving differential inclusions and systems with control or state constraints. When state constraints are present, additional transversality and sensitivity conditions involving the adjoint (or co-state) arc become necessary. In smooth settings, these conditions express orthogonality of the adjoint endpoint to the tangent hyperplane of the constraint boundary. The adjoint variable also encodes sensitivity information on the dependence of the value function on the initial state. However, the presence of state constraints complicates the application of the dynamic programming principle and the analysis of the HJB equation, since differentiability or even continuity of the value function may fail. In such cases, generalized solution concepts, such as viscosity solutions introduced by Crandall, Evans, and Lions, are required; yet, under state constraints, one can generally only expect lower semicontinuity.

Because of the widespread use of infinite horizon models in applied sciences, the study of nonautonomous control systems subject to nonsmooth and unbounded state constraints is of fundamental importance. The aim of this book is to provide a concise introduction to nonsmooth analysis techniques suitable for such problems, where both the data and the constraint structure lack smoothness. While optimal synthesis for finite horizon problems of Mayer or Bolza type is well known, these results do not extend directly to the infinite horizon case, since truncating an infinite horizon trajectory does not yield an optimal solution for the finite horizon problem. For infinite horizon systems with nonsmooth dynamics and state constraints, necessary conditions are formulated in terms of the maximum principle, extended Euler-Lagrange inclusions, and partial sensitivity relations. A normal maximum principle, together with partial and full sensitivity relations and a transversality condition at the initial time, is established under weak regularity assumptions. A key aspect of the analysis concerns the regularity of the value function, since understanding its properties refines optimality conditions and clarifies the stability of optimal trajectories. The study includes sufficient conditions ensuring Lipschitz continuity of the value function in the presence of time-dependent dynamics and a nonautonomous Lagrangian. Lipschitz regularity follows from estimates on the distance between a trajectory and the set of feasible trajectories, obtained under a uniform inward-pointing condition on the state constraints. This guarantees Lipschitz dependence of feasible trajectories on the initial state. While control constraints are well understood, state constraints introduce significant analytical difficulties: small perturbations in initial conditions or control inputs may lead to constraint violations, and the value function can become unbounded or discontinuous. These issues highlight the need for a deeper understanding of the regularity and weak-solution properties of HJB equations, which are crucial for the robustness and applicability of infinite horizon control methods in practical settings.

The book is organized as follows. Chapter 1 recalls essential results from nonsmooth analysis. Chapter 2 studies the Lipschitz continuity of the value function and neighboring feasible trajectories. Chapter 3 develops the maximum principle, together with sensitivity and transversality conditions for the co-state. Finally, Chap. 4 focuses on HJB equations with time-measurable data and introduces a new concept of weak epigraphical solution designed to handle discontinuities.

Rome, Italy
October 2025

Vincenzo Basco

Competing Interests The author has no competing interests to declare that are relevant to the content of this manuscript.

Contents

Notation

$\mathbb{N}^+$; $\mathbb{R}_t^+$	$\{n \in \mathbb{N} \mid 0 < n\}$; $\{s \in \mathbb{R} \mid t \leq s\}$
$a \wedge b$; $a \vee b$	$\min\{a, b\}$; $\max\{a, b\}$
$\langle ., . \rangle$; $\lvert . \rvert$	Euclidean scalar product; Euclidean norm
$\mathrm{dist}(X, Y)$; $d_Y(x)$	$\inf\{\lvert x - y\rvert \mid x \in X, y \in Y\}$; $\mathrm{dist}(\{x\}, Y)$
$d_{\mathscr{H}}(X, Y)$	Hausdorff distance, i.e., $\sup_{y\in Y} d_X(y) \vee \sup_{x\in X} d_Y(x)$
$B(x, r)$	Euclidean ball centered at x of radius r
$\mathbb{B}$; $\mathbb{S}$	$B(0, 1)$; $\{x \in B(0, 1) \mid \lvert x\rvert = 1\}$
$\mu_{\mathscr{L}}$	Lebesgue measure
A^c	Complement of the set A
bdr A	Boundary of the set A
int A	Interior of the set A
cl A	Closure of the set A
co A	Convex hull of the set A
χ_A	Indicator function of the set A, i.e., $\chi_A(x) = 1$ if $x \in A$ and $\chi_A(x) = 0$ elsewhere
∇f; $\nabla_x f$	Gradient of f; Gradient of f wrt x
$\mathrm{d}f$; $\mathrm{d}_x f$	Differential of f; Differential of f wrt x
$L^p(I; A)$	Space of all $f : I \to A$ Lebesgue measurable, a.e. equal, endowed with $\lVert f\rVert_{p,I}^p = \int_I \lvert f(s)\rvert^p ds$, $p \geq 1$
$L^p_{\mathrm{loc}}(I; A)$	Set of all $f : I \to A$ Lebesgue measurable, a.e. equal, $f\vert_J \in L^p(J; A)$ for any $J \subset I$ compact interval
$L^\infty(I; A)$	Space of all $f : I \to A$ Lebesgue measurable, a.e. equal, endowed with $\lVert f\rVert_{\infty,I} = \text{ess-sup}\{\lvert f(s)\rvert \mid s \in I\}$
$W^{1,1}(I; A)$	Space of all $f : I = [a, b] \to A$ absolutely continuous endowed with $\lVert f\rVert_{W^{1,1},I} = \lvert f(a)\rvert + \int_a^b \lvert f'(s)\rvert ds$
$W^{1,1}_{\mathrm{loc}}(I; A)$	Set of all $f : I \to A$ Lebesgue measurable, $f\vert_J \in W^{1,1}(J; A)$ for any $J \subset I$ compact interval
$C(I; A)$	Space of all $f : I \to A$ continuous endowed with $\lVert f\rVert_{\infty,I} = \sup\{\lvert f(s)\rvert \mid s \in I\}$
$C(I; A)^*$	Dual of $C(I; A)$

$\theta_f(\sigma)$	$\sup\left\{\int_J \lvert f(s)\rvert\, ds \,\middle|\, J \subset \mathbb{R}_0^+,\ \mu_{\mathscr{L}}(J) \leqslant \sigma\right\}$ for any $f \in L^1_{\mathrm{loc}}(\mathbb{R}_0^+; \mathbb{R}_0^+)$
$\mathcal{L}_{\mathrm{loc}}$	Set of all $f \in L^1_{\mathrm{loc}}(\mathbb{R}_0^+; \mathbb{R}_0^+)$ such that $\lim_{\sigma \to 0} \theta_f(\sigma) = 0$
A^-; A^+	Negative polar of the set A, i.e., $\{p \mid \langle p, \alpha\rangle \leq 0 \text{ for all } \alpha \in A\}$; $-A^-$

Chapter 1
Preliminaries on Nonsmooth Analysis

Abstract In this chapter, we revisit some definitions and results in nonsmooth analysis that are instrumental in subsequent chapters, introducing some basic mathematical concepts utilized in this book. For a more comprehensive and thorough treatment, we refer readers to the monographs by Clarke (Optimization and Nonsmooth Analysis, Volume 5 of Classics in Applied Mathematics, 2nd edn. Society for Industrial and Applied Mathematics (SIAM). Philadelphia, PA, 1990), Rockafellar and Wets (Variational Analysis, Volume 317 of Grundlehren der Mathematischen Wissenschaften [Fundamental Principles of Mathematical Sciences]. Springer, Berlin, 1998), Vinter (Optimal Control. Birkhäuser, Boston, MA, 2000) and the references at the end of this section, where they can delve deeper into the topics briefly touched upon here.

Keywords Normal cones · Tangent cones · Set-valued maps · Regularity notions

1.1 Normal and Tangent Cones

The geometric formulations of normal and tangent cones play a fundamental role in the analysis of constrained optimization problems and differential inclusions. These cones provide essential tools for characterizing feasible directions and optimality conditions in nonsmooth settings. This section establishes the basic definitions and properties of proximal, regular, and limiting normal cones, along with their corresponding tangent cone counterparts, which will be extensively used throughout the subsequent analysis.

Let $E \subset \mathbb{R}^k$ be nonempty and $x \in \operatorname{cl} E$. The *proximal/regular/limiting normal cones* to E at x are defined, respectively, by[1]

$$N_E^P(x) := \left\{ p \in \mathbb{R}^k \mid \exists M \geq 0,\ \langle p, y - x \rangle \leq M |y - x|^2 \, \forall y \in E \right\},$$

$$\widehat{N}_E(x) := \{ p \in \mathbb{R}^k \mid \limsup_{\substack{y \to x \\ E}} \frac{\langle p, y - x \rangle}{|y - x|} \leq 0 \},$$

[1] We write $x_i \underset{E}{\to} x$ for $x_i \to x$ and $x_i \in E$ for all $i \in \mathbb{N}$.

V. Basco, *Nonsmooth Constrained Optimal Control on Infinite Horizon*,
SpringerBriefs in Mathematics, https://doi.org/10.1007/978-3-032-08938-0_1

$$N_E(x) := \{p \in \mathbb{R}^k \mid \exists x_i \underset{E}{\to} x, \exists p_i \to p, p_i \in N_E^P(x_i) \forall i \in \mathbb{N}\}.$$

From the definitions above it follows the following factorization property: for any closed subsets $E_1 \subset \mathbb{R}^m$ and $E_2 \subset \mathbb{R}^k$ and any point $(x_1, x_2) \in E_1 \times E_2$,

$$\begin{aligned} N^P_{E_1 \times E_2}(x_1, x_2) &= N^P_{E_1}(x_1) \times N^P_{E_2}(x_2), \\ \widehat{N}_{E_1 \times E_2}(x_1, x_2) &= \widehat{N}_{E_1}(x_1) \times \widehat{N}_{E_2}(x_2), \\ N_{E_1 \times E_2}(x_1, x_2) &= N_{E_1}(x_1) \times N_{E_2}(x_2). \end{aligned}$$

Lemma 1.1.1 *Let $E \subset \mathbb{R}^k$ be closed nonempty and $x \in E$. Then:*

(i) *$p \in N_E(x)$ if and only if there exist sequences $x_i \to x$ and $p_i \to p$ such that $x_i \in E$ and $p_i \in \widehat{N}_E(x_i)$ for all $i \in \mathbb{N}$;*
(ii) *$N_E^P(x)$, $\widehat{N}_E(x)$, $N_E(x)$ are cones and $0 \in N_E^P(x) \subset \widehat{N}_E(x) \subset N_E(x)$;*
(iii) *if $x \in \operatorname{bdr} E$ then $(N_E(x) \setminus \{0\}) \cap \mathbb{R}^k \neq \emptyset$;*
(iv) *if $x \in \operatorname{int} E$ then $N_E(x) = \{0\}$;*
(v) *$N_E^P(x)$ is convex and $\widehat{N}_E(x)$ is closed and convex. The elements $p \in N_E^P(x)$ are also called proximal normals and there exists $\lambda = \lambda(p) > 0$ such that $d_E(x + \lambda p) = \lambda |p|$, i.e.,*

$$\operatorname{int} B(x + \lambda p, \lambda |p|) \subset E^c; \tag{1.1}$$

(vi) *if E is convex then $N_E^P(x) = \widehat{N}_E(x) = N_E(x)$, $N_E(x)$ is also called normal cone, and it holds*

$$p \in N_E(x) \Longleftrightarrow \langle p, y - x \rangle \le 0 \ \forall y \in E; \tag{1.2}$$

(vii) *for any $y_i \underset{E}{\to} y$ and any $p_i \to p$ such that $p_i \in N_E(y_i)$ for all $i \in \mathbb{N}$ then $p \in N_E(y)$.*

Proof See [9, Chap. 4, Sect. 2] and [8]. □

The *Bouligad/Clarke tangent cones* to E at x are defined, respectively, by

$$\begin{aligned} T_E(x) &:= \{\xi \in \mathbb{R}^k \mid \exists t_i \to 0+, \exists \xi_i \to \xi, \ x + t_i \xi_i \in E \, \forall i \in \mathbb{N}\}, \\ T_E^C(x) &:= \{\xi \in \mathbb{R}^k \mid \forall x_i \to_E x, \forall t_i \to 0+, \exists \xi_i \to \xi, \ x_i + t_i \xi_i \in E \, \forall i \in \mathbb{N}\}. \end{aligned}$$

We denote by

$$N_E^C(x) := T_E^C(x)^-$$

the *Clarke normal cone* to E at x.

Lemma 1.1.2 *Let $E \subset \mathbb{R}^n$ be closed nonempty and $x \in E$. Then:*

(i) *$T_E(x)$ and $T_E^C(x)$ are closed and $T_E^C(x)$ is convex;*

(ii) $\xi \in T_E(x)$ *if and only if* $\liminf_{h\to 0+} \dfrac{d_E(x+h\xi)}{h} = 0$;

(iii) $T_E^C(x) = N_E(x)^- \subset T_E(x)$;

(iv) $\xi \in \operatorname{int} T_E(x)$ *if any only if there exist* $\varepsilon > 0$, $\delta > 0$, *and* $\lambda > 0$ *such that* $\tilde{x} + t\tilde{\xi} \in E$ *for all* $\tilde{x} \in E \cap (x + \delta\mathbb{B})$, $t \in [0, \lambda]$, *and* $\tilde{\xi} \in (\xi + \varepsilon\mathbb{B})$;

(v) $T_E(x)^- = \widehat{N}_E(x)$;

(vi) $\operatorname{cl}\operatorname{co} N_E(x) = N_E^C(x)$.

Proof See [9, Proposition 4.10.3], [9, Chap. 4, Sect. 2] and [8, Chap. 6]. □

1.2 Subdifferentials

The notion of subdifferential extends the concept of derivative to nonsmooth functions, providing a tool for analyzing optimization problems where classical differentiability fails. In the context of optimal control under state constraints, subdifferentials are crucial for characterizing optimality conditions and studying the regularity properties of value functions. This section presents the fundamental notions and properties of subdifferentials, including contingent derivatives.

Let $\varphi : \mathbb{R}^k \to [-\infty, +\infty]$ be an extended real function.[2] The *domain*, *epigraph*, and *hypograph* of φ are defined, respectively, by

$$\begin{aligned} \operatorname{dom}\varphi &:= \{x \in \mathbb{R}^k \mid \varphi(x) < +\infty\}, \\ \operatorname{epi}\varphi &:= \{(x,\lambda) \in \mathbb{R}^k \times \mathbb{R} \mid \varphi(x) \le \lambda\}, \\ \operatorname{hypo}\varphi &:= \{(x,\lambda) \in \mathbb{R}^k \times \mathbb{R} \mid \varphi(x) \ge \lambda\}. \end{aligned}$$

The *subdifferential* and the *limiting subdifferential/superdifferential* of φ at $x \in \operatorname{dom}\varphi$ are defined, respectively, by

$$\begin{aligned} \widehat{\partial}\varphi(x) &:= \left\{\xi \in \mathbb{R}^k \mid (\xi,-1) \in \widehat{N}_{\operatorname{epi}\varphi}(x,\varphi(x))\right\}, \\ \partial\varphi(x) &:= \left\{\xi \in \mathbb{R}^k \mid (\xi,-1) \in N_{\operatorname{epi}\varphi}(x,\varphi(x))\right\}, \\ \partial^+\varphi(x) &:= \left\{\xi \in \mathbb{R}^k \mid (-\xi,1) \in N_{\operatorname{hypo}\varphi}(x,\varphi(x))\right\}. \end{aligned}$$

The set $\partial\varphi(x)$ is closed whenever φ is lower semicontinuous (briefly, *l.s.c.*), proper, and $x \in \operatorname{dom}\varphi$, because of the definition of limiting normal cone and since $(x,\lambda) \rightsquigarrow N_{\operatorname{epi}\varphi}(x,\lambda)$ has closed graph as map from $\operatorname{epi}\varphi$ to $\mathbb{R}^k \times \mathbb{R}$ (cfr. Lemma 1.1.1).

The *Clarke generalized subdifferential* of φ at x is defined as

$$\partial^C\varphi(x) := \{\xi \in \mathbb{R}^k \mid \limsup_{\substack{y\to x\\ t\to 0+}} \frac{\varphi(y+tv)-\varphi(y)}{t} \ge \langle \xi, v\rangle \ \forall v \in \mathbb{R}^k\}.$$

---

[2] We recall that φ is said to be measurable if $\varphi^{-1}(\{+\infty\})$, $\varphi^{-1}(\{-\infty\})$, and $\varphi^{-1}(I)$ are measurable for any Borel subset $I \subset \mathbb{R}$.

If φ is Lipschitz continuous on a neighborhood of $x \in \operatorname{dom}\varphi$, then (see [9, Proposition 4.7.6] and [8, Theorem 8.49]) the sets $\partial\varphi(x)$, $\partial^+\varphi(x)$ are nonempty, the set $\partial^C\varphi(x)$ is closed, and for any subset $\Omega \subset \mathbb{R}^k$ of null Lebesgue measure

$$\partial^C\varphi(x) = \operatorname{co}\partial\varphi(x) = \operatorname{co}\{\xi \in \mathbb{R}^k \mid \exists x_i \xrightarrow[\mathbb{R}^k\setminus\Omega]{} x, \exists\nabla\varphi(x_i), \nabla\varphi(x_i) \to \xi\}. \tag{1.3}$$

Lemma 1.2.1 *Let $\varphi(.) = d_C(.)$ with $C \subset \mathbb{R}^n$ closed nonempty and $\bar{x} \in C$. Then:*

$$\partial\varphi(\bar{x}) = N_C(\bar{x}) \cap \mathbb{B}, \quad \widehat{\partial}\varphi(\bar{x}) = \widehat{N}_C(\bar{x}) \cap \mathbb{B}, \quad d\varphi(\bar{x})(w) = d_{T_C(\bar{x})}(w).$$

Moreover, for any $\bar{x} \notin C$:

$$\partial\varphi(\bar{x}) = \frac{\bar{x} - \Pi_C(\bar{x})}{d_C(\bar{x})},$$

$$\widehat{\partial}\varphi(\bar{x}) = \begin{cases} \left\{\frac{\bar{x}-y}{d_C(\bar{x})}\right\} & \text{if } \Pi_C(\bar{x}) = \{y\}, \\ \emptyset & \text{otherwise}, \end{cases}$$

$$d\varphi(\bar{x})(w) = \min_{y\in\Pi_C(\bar{x})} \frac{\langle\bar{x} - y, w\rangle}{d_C(\bar{x})},$$

where

$$\Pi_C(\bar{x}) := \{x \in C \mid |\bar{x} - x| = d_C(\bar{x})\} \tag{1.4}$$

is the projection operator onto C of $\bar{x}$.

Proof See [8, pp. 340–341]. □

The *contingent epiderivative/hypoderivative*, in direction $u \in \mathbb{R}^k$, of φ at $x \in \operatorname{dom}\varphi$ are, respectively, defined by

$$D_\uparrow\varphi(x)(u) := \liminf_{\substack{h\to 0+ \\ \tilde{u}\to u}} \frac{\varphi(x + h\tilde{u}) - \varphi(x)}{h},$$

$$D_\downarrow\varphi(x)(u) := \limsup_{\substack{h\to 0+ \\ \tilde{u}\to u}} \frac{\varphi(x + h\tilde{u}) - \varphi(x)}{h}.$$

The *Fréchet subdifferential/superdifferential* of φ at $x \in \operatorname{dom}\varphi$ are, respectively, defined by

$$\partial_-\varphi(x) := \{p \in \mathbb{R}^k \mid \liminf_{y\to x} \frac{\varphi(y) - \varphi(x) - \langle p, y - x\rangle}{|y - x|} \geq 0\},$$

$$\partial_+\varphi(x) := \{p \in \mathbb{R}^k \mid \limsup_{y\to x} \frac{\varphi(y) - \varphi(x) - \langle p, y - x\rangle}{|y - x|} \leq 0\}.$$

From the definitions of contingent derivatives and Boulingad tangent cone (cfr. also [1, pp. 188–191]), we have

$$\text{epi } D_{\uparrow}\varphi(x) = T_{\text{epi}\,\varphi}(x, \varphi(x)) \;\&\; \text{hypo } D_{\downarrow}\varphi(x) = T_{\text{hypo}\,\varphi}(x, \varphi(x_0)). \quad (1.5)$$

Lemma 1.2.2 *For any $x \in \text{dom}\,\varphi$:*

(i) if φ is convex and proper, then

$$\partial_+\varphi(x) = \partial\varphi(x) = \{p \in \mathbb{R}^k \mid \varphi(y) \geq \varphi(x) + \langle p, y - x\rangle \ \forall y \in \mathbb{R}^k\}$$

and $\partial\varphi(x)$ is called subdifferential (in the sense of convex analysis) of φ at x;

(ii) if φ is lower semicontinuous, for any $(p, 0) \in N_{\text{epi}\,\varphi}(x, \varphi(x))$ there exist two sequences $x_i \in \text{dom}\,\varphi$ and $(p_i, q_i) \in N_{\text{epi}\,\varphi}(x_i, \varphi(x_i))$ such that $q_i < 0$ for all $i \in \mathbb{N}$ and $(x_i, \varphi(x_i)) \to (x, \varphi(x))$, $(p_i, q_i) \to (p, 0)$.

Proof We refer to [7, Theorem 1] and [8, Proposition 8.12]. □

The *Fenchel transform* (or *conjugate*) of an extended real function $\varphi : \mathbb{R}^k \to [-\infty, +\infty]$, written φ^*, is the extended real function $\varphi^* : \mathbb{R}^k \to [-\infty, +\infty]$ defined by

$$\varphi^*(v) := \sup_{p\in\mathbb{R}^k}\{\langle v, p\rangle - \varphi(p)\}.$$

Lemma 1.2.3 *Assume that φ is proper, lower semicontinuous, and convex.*

Then φ^* is a proper lower semicontinuous convex function, $(\varphi^*)^* = \varphi$, $\text{dom}\,\varphi^*$ is convex, and for all $p, v \in \mathbb{R}^k$ it holds that:

$$p \in \partial\varphi^*(v) \iff v \in \partial\varphi(p) \iff \varphi(p) + \varphi^*(v) = \langle v, p\rangle.$$

Proof See [8, Theorems 11.1, 11.3]. □

1.3 Set-Valued Maps

Set-valued maps arise naturally in the study of differential inclusions and constrained optimization problems, where the dynamics or constraint sets may exhibit multivalued behavior. Understanding the continuity and regularity properties of such maps is essential for establishing existence results and feasibility of solutions to set-valued inclusions.

Let $\mathscr{X} \subset \mathbb{R}^m$ and $\mathscr{Y} \subset \mathbb{R}^k$ be two nonempty sets. The graph and the domain of a set-valued map $\Phi : \mathscr{X} \rightsquigarrow \mathscr{Y}$ are defined, rispectively, by

$$\text{graph}\,\Phi := \{(x, y) \in \mathscr{X} \times \mathscr{Y} \mid y \in \Phi(x)\},$$
$$\text{dom}\,\Phi := \{x \in \mathscr{X} \mid \Phi(x) \neq \emptyset\}.$$

Let $\tilde{x} \in \mathscr{X}$. The sets

$$\operatorname{Lim\,inf}_{x \xrightarrow[\mathscr{X}]{} \tilde{x}} \Phi(x) := \{\xi \in \mathbb{R}^k \mid \forall x_i \xrightarrow[\mathscr{X}]{} \tilde{x},\ \exists \xi_i \to \xi,\ \xi_i \in \Phi(x_i)\ \forall i \in \mathbb{N}\},$$
$$\operatorname{Lim\,sup}_{x \xrightarrow[\mathscr{X}]{} \tilde{x}} \Phi(x) := \{\xi \in \mathbb{R}^k \mid \exists x_i \xrightarrow[\mathscr{X}]{} \tilde{x},\ \exists \xi_i \to \xi,\ \xi_i \in \Phi(x_i)\ \forall i \in \mathbb{N}\},$$

are called, respectively, the *lower/upper limits* (in the Kuratowski-Painlevé sense) of $\Phi : \mathscr{X} \rightsquigarrow \mathscr{Y}$ at $\tilde{x}$. The above definition apply also for any sequence of subsets $(\mathscr{A}_i)_{i \in \mathbb{N}^+} \subset \mathbb{R}^k$ where

$$\operatorname*{Lim\,inf}_{i \to +\infty} \mathscr{A}_i := \operatorname*{Lim\,inf}_{x \xrightarrow[\mathscr{X}]{} 0} \Phi(x),$$
$$\operatorname*{Lim\,sup}_{i \to +\infty} \mathscr{A}_i := \operatorname*{Lim\,sup}_{x \xrightarrow[\mathscr{X}]{} 0} \Phi(x),$$

and $\Phi(x) = \mathscr{A}_{1/x}$, $x \in \mathscr{X} = \{\frac{1}{i} \mid i \in \mathbb{N}^+\}$.

Observe that the lower and upper limits are closed, possibly empty, and verify

$$\operatorname*{Lim\,inf}_{x \xrightarrow[\mathscr{X}]{} \tilde{x}} \Phi(x) \subset \operatorname*{Lim\,sup}_{x \xrightarrow[\mathscr{X}]{} \tilde{x}} \Phi(x).$$

A set-valued map $\Phi : \mathscr{X} \rightsquigarrow \mathscr{Y}$ is said to be:

- *upper semicontinuous at* $\tilde{x} \in \mathscr{X}$ if for any $\varepsilon > 0$ there exists $\delta > 0$ such that $\Phi(x) \subset \Phi(\tilde{x}) + \varepsilon \mathbb{B}$ for all $x \in B(\tilde{x}, \delta) \cap \mathscr{X}$. If Φ is upper semicontinuous at every $\tilde{x} \in \mathscr{X}$ then it is said to be *upper semicontinuous*;
- *lower semicontinuous at* $\tilde{x} \in \mathbb{R}^m$ if $\operatorname{Lim\,inf}_{x \to_{\mathscr{X}} \tilde{x}} \Phi(x) \subset \Phi(\tilde{x})$. If Φ is lower semicontinuous a every $\tilde{x} \in \mathscr{X}$ then it is said to be *lower semicontinuous*;
- *continuous at* $x \in \mathscr{X}$ if it is lower and upper semicontinuous at x and it is *continuous* if it is continuous at each point $x \in \mathscr{X}$;
- *L-Lipschitz continuous* (or *Lipschitz continuous*), for some $L \geq 0$, if $\Phi(x) \subset \Phi(\tilde{x}) + L|x - \tilde{x}|\mathbb{B}$ for all $x, \tilde{x} \in \mathscr{X}$.

Below, there is a version of Scorza-Dragoni Theorem.

Lemma 1.3.1 *Let X be a metric space and $\Phi : [0, T] \times X \rightsquigarrow \mathbb{R}^k$ be a set-valued map with nonempty convex closed values and:*

- *$\Phi(t, .)$ is upper semicontinuous for almost all $t \in [0, T]$;*
- *$\sup_{v \in \Phi(t,z)} |v| \leqslant \mu(t)$ for almost all $t \in [0, T]$ and every $z \in X$ and a suitable measurable function $\mu : [0, T] \to \mathbb{R}$.*

Then there exists a set-valued map $\Phi_\sharp : [0, T] \times X \rightsquigarrow \mathbb{R}^k$ with nonempty closed convex values satisfying the following properties:

(i) $\Phi_\sharp(t, z) \subset \Phi(t, z)$ for almost all $t \in [0, T]$ and for all $z \in X$;

(ii) *For every measurable set $C \subset [0, T]$ and every $y : C \to X$, $z : C \to \mathbb{R}^k$ measurable maps such that $z(t) \in \Phi(t, y(t))$ a.e. in C we have $z(t) \in \Phi_\sharp(t, y(t))$ a.e. in C;*

(iii) *For any $\delta > 0$ there is a closed set $C^\delta \subset [0, T]$ such that $\mu_{\mathcal{L}}([0, T] \setminus C^\delta) < \delta$ and $\Phi_\sharp$ is an upper semicontinuous map on $C^\delta \times X$.*

Proof See [6]. □

Lemma 1.3.2 *Consider a set valued map $\Phi : \mathbb{R}^k \rightsquigarrow \mathbb{R}^k$ and assume that for some $y_0 \in \mathbb{R}^k$ and $\varepsilon > 0$, $c \geq 0$, Φ is c-Lipschitz continuous on the ball $y_0 + \varepsilon\mathbb{B}$. Then, for every $\bar{y} \in \mathbb{R}^k$ satisfying $|\bar{y} - y_0| < \varepsilon$ and any $\bar{z} \in \Phi(\bar{y})$, $(r, q) \in \mathbb{R}^k \times \mathbb{R}^k$, the following implication holds true*

$$(r, q) \in N_{\text{graph}\,\Phi}(\bar{y}, \bar{z}) \Longrightarrow |r| \leq c|q|.$$

Proof See [3, Proposition 1]. □

For any $\tau \in I \subset \mathbb{R}$ and any $y \in \mathbb{R}^k$ we denote by $\mathrm{Ker}\,(\Phi, \tau, y)$ the (possibly empty) family of locally absolutely continuous function $x : I \to \mathbb{R}^k$ solutions of the problem

$$\begin{cases} x'(t) \in \Phi(t, x(t)) & \text{a.e. } t \in I, \\ x(\tau) = y. \end{cases}$$

We conclude this section with the following relaxation result.

Lemma 1.3.3 *Consider a measurable set-valued map $\Phi : [S, T] \times \mathbb{R}^k \rightsquigarrow \mathbb{R}^k$ with closed and nonempty values. Assume that there exist $\varphi, \psi \in L^1([S, T]; \mathbb{R})$ such that:*

$$\Phi(t, x) \subset \Phi(t, y) + \varphi(t)\mathbb{B} \qquad \forall t \in [S, T], \forall x, y \in \mathbb{R}^k,$$

$$\Phi(t, x) \subset \psi(t)\mathbb{B} \qquad \forall (t, x) \in [S, T] \times \mathbb{R}^k.$$

Let $\bar{x} \in \mathbb{R}^k$ and take any $x \in \mathrm{Ker}(\mathrm{co}\,\Phi, S, \bar{x})$ and any $\varepsilon > 0$. Then there exists $y \in \mathrm{Ker}(\Phi, S, \bar{x})$ such that $\|y - x\|_{\infty,[S,T]} < \varepsilon$.

Proof See [9, Theorem 2.7.2, p. 96]. □

1.3.1 The Steiner Map

The Steiner map provides a canonical way to associate points in space with convex sets, playing an important role in the geometric analysis of optimization problems. This subsection introduces the Steiner map and establishes its key properties, particularly its Lipschitz continuity characteristics that will prove useful in representation of Hamiltonians.

Consider the map

$$S_k : \{J \subset \mathbb{R}^k \mid J \text{ convex nonempty bounded}\} \to \mathbb{R}^k$$

defined by

$$S_k(J) := \frac{1}{\mu_{\mathscr{L}}(\mathbb{B})} \int_{\mathbb{B}} \Pi_{\partial\sigma_J(p)}(0)\, \mu_{\mathscr{L}}(dp),$$

where: for any K nonempty convex, $\Pi_K(0)$ stands for the projection of $0 \in \mathbb{R}^k$ onto K, i.e., the element in K with minimal norm; $\sigma_J(.)$ denotes the support function of J, that is

$$\sigma_J(p) := \sup_{q \in J} \langle p, q \rangle.$$

The function $S_k(.)$ is called *Steiner map (or Steiner selection).*

Lemma 1.3.4 *The function $S_k(.)$ is k-Lipschitz continuous (wrt the Hausdorff distance $d_{\mathscr{H}}$) on the family of all convex nonempty bounded subset of $\mathbb{R}^k$ and satisfies*

$$S_k(J) \in J \qquad \forall J \subset \mathbb{R}^k \text{ convex nonempty bounded.} \tag{1.6}$$

Proof We notice that (1.6) follows immediately from the properties of Fenchel transform and the definition of subdifferential. Indeed, fix $J \subset \mathbb{R}^k$ convex nonempty bounded and let $p \in \mathbb{B}$. Define $\psi(.) = \psi_p(.) := \sigma_J(. + p)$. The function ψ is proper[3] convex. From Lemma 1.2.3, it follows that

$$\partial\sigma_J(p) = \partial\psi(0) = \arg\min \psi^*. \tag{1.7}$$

So, $\psi^*(q) = \sup_{y \in \mathbb{R}^k}\{\langle y, q\rangle - \sigma_J(y + p)\} = -\langle p, q\rangle$, if $q \in J$, and $+\infty$ otherwise. From 1.7, we have $\partial\sigma_J(p) = \arg\max_{q \in J}\langle p, q\rangle$, and, by arbitrariness of p, we conclude $\Pi_{\partial\sigma_J(p)}(0) \in J$ for all $p \in \mathbb{B}$. Thus, since J is convex and from the definition of the Bochner integral, we get (1.6).

For a proof of the Lipschitz continuity we refer the reader to [10, Appendix]. □

1.3.2 Regularity Notions

Different notions of regularity for set-valued maps are in the following introduced to address the analysis of control problems involving multivalued differential inclusions. The next result provide a characterization of continuity.

Lemma 1.3.5 *Let $X \subset \mathbb{R}^m$ be a closed nonempty set. Consider a set-valued map $\Phi : X \rightsquigarrow \mathbb{R}^k$ and let $\bar{x} \in X \cap \operatorname{dom}\Phi$. Then Φ is continuous at $\bar{x}$ if and only if the function $y \mapsto d_{\Phi(y)}(x)$ is continuous at $\bar{x}$ for every $x \in \mathbb{R}^m$.*

Proof See [5, Chap. 3]). □

[3] i.e., $\psi(.) \neq -\infty$ and there exits x with $\psi(x) < +\infty$.

We say that a set-valued map $\Phi : [S, T] \rightsquigarrow \mathbb{R}^k$ is of *bounded variations* if Φ takes nonempty closed images and for any $\mathscr{K} \subset \mathbb{R}^k$

$$\sup \sum_{i=1}^{m-1} exc(\Phi(t_{i+1}) \cap \mathscr{K} | \Phi(t_i)) \vee exc(\Phi(t_i) \cap \mathscr{K} | \Phi(t_{i+1})) < +\infty$$

where the supremum is taken over all finite partition $a = t_1 < t_2 < ... < t_{m-1} < t_m = b$ and the *excess* of A given B is defined by

$$exc(A|B) := \sup\{d_B(a) \,|\, a \in A\} \in [0, +\infty]. \tag{1.8}$$

It is possible to show that $exc(A|B) = \inf\{r > 0 \,|\, B \subset A + r\mathbb{B}\}$ for all $A, B \subset \mathbb{R}^k$, where the infimum over an empty set is defined as $+\infty$.

We say that a set-valued map $\Phi : \mathbb{R}_S^+ \rightsquigarrow \mathbb{R}^k$ is of *locally bounded variations* (briefly, *LBV*) if $\Phi|_{[S,T]}$ is of bounded variations for any $T > S$.

A set-valued map $\Phi : [S, T] \rightsquigarrow \mathbb{R}^k$ is *absolutely continuous* if it takes nonempty closed images and for every $\varepsilon > 0$, and any compact subset $\mathscr{K} \subset \mathbb{R}^k$, there exists $\delta > 0$ such that for any finite partition $S \le t_1 < \tau_1 \le t_2 < \tau_2 \le ... \le t_m < \tau_m \le T$ of $[S, T]$,

$$\sum_{i=1}^{m} (\tau_i - t_i) < \delta \Longrightarrow \sum_{i=1}^{m} exc(\Phi(\tau_i) \cap \mathscr{K} | \Phi(t_i)) \vee exc(\Phi(t_i) \cap \mathscr{K} | \Phi(\tau_i)) < \varepsilon.$$

If $I = \mathbb{R}_S^+$, we say that $\Phi : I \rightsquigarrow \mathbb{R}^k$ is *locally absolutely continuous* (briefly, *l.a.c.*) if $\Phi|_{[S,T]}$ is absolutely continuous for any $T > S$.

Let $I \subset \mathbb{R}$ be a nonempty interval, $\Phi : \mathrm{cl}\, I \times \mathbb{R}^k \rightsquigarrow \mathbb{R}^k$ and $\Omega : \mathbb{R} \rightsquigarrow \mathbb{R}^k$ be set-valued maps with nonempty values, and $\gamma \in L^1_{\mathrm{loc}}(\mathrm{cl}\, I; \mathbb{R}_0^+)$. We say that Φ is *γ-left absolutely continuous, uniformly wrt* Ω, if

$$\Phi(s, x) \subset \Phi(t, x) + \int_s^t \gamma(\tau)\, d\tau \mathbb{B} \quad \forall s, t \in \mathrm{cl}\, I \,:\, s < t,\ \forall x \in \cup_{\tau \in [s,t]} \Omega(\tau). \tag{1.9}$$

If Φ does not depends explicitly from x, in that case we simply say that Φ is *γ-left absolutely continuous*. If $\mathrm{cl}\, I = [S, T]$, then we have the following characterization of uniform absolute continuity from the left: Φ is γ-left absolutely continuous, uniformly wrt Ω, if and only if for every $\varepsilon > 0$ there exists $\delta > 0$ such that for any finite partition $S \le t_1 < \tau_1 \le t_2 < \tau_2 \le ... \le t_m < \tau_m \le T$ of $[S, T]$,

$$\sum_{i=1}^{m} (\tau_i - t_i) < \delta \Longrightarrow \sum_{i=1}^{m} exc(\Phi(t_i, x) | \Phi(\tau_i, x)) < \varepsilon \quad \forall\, x \in \cup_{\tau \in [S,T]} \Omega(\tau).$$

We say that Φ has a *sub-linear growth* (in x) if, for some $c \in L^1_{\mathrm{loc}}(\mathrm{cl}\, I; \mathbb{R}_0^+)$, $\sup_{\phi \in \Phi(t,x)} |\phi| \le c(t)(1 + |x|)$ for a.e. $t \in \mathrm{cl}\, I$ and for all $x \in \mathbb{R}^k$.

1.3.3 Further Results

The theoretical framework developed in this chapter relies on several deep results from set-valued analysis and measurable selection theory that, while fundamental to our approach, would require extensive development if presented in full detail. Rather than interrupt the flow of our main arguments, we collect here the essential theorems that underpin our analysis: from the measurability characterizations that ensure our differential inclusions are well-posed, to the selection theorems that guarantee the existence of measurable controls, and the continuity results that enable our regularity analysis of the value function. These classical results, established by pioneers like Michael, Filippov, and others, represent some of the most fundamental tools in modern nonsmooth (and variational) analysis and are indispensable for rigorous treatment of infinite horizon optimal control problems with state constraints:

- Characterization of Measurability of Set-Valued Maps ([2, Chap. III, p. 59; Theorem III.30 p. 80]);
- Measurable Selection Theorem ([1, Theorem 1, p. 90; Corollary 1, p. 91], [9, Theorem 2.3.12, p. 71]);
- Continuous Selection Theorem (known also as Michael's Theorem, [1, Theorem 1, p. 82]);
- Generalized Filippov Existence Theorem ([9, Theorem 2.4.3, p. 78]);
- Mean Value Theorem ([4, Theorem 2.3.7, p. 41], [9, Theorem 4.5.3, p. 149]).

References

1. Aubin, J.-P., Cellina, A.: Differential Inclusions: Set-valued Maps and Viability Theory, vol. 264. Springer Science & Business Media (2012)
2. Castaing, C., Valadier, M.: Convex Analysis and Measurable Multifunctions. Lecture Notes in Mathematics. Springer, Berlin, Heidelberg (1977)
3. Clarke, F.: A general theorem on necessary conditions in optimal control. Discrete Continuous Dyn. Syst. **29**(2), 485–503 (2011)
4. Clarke, F.H.: Optimization and Nonsmooth Analysis, Volume 5 of Classics in Applied Mathematics, 2nd edn. Society for Industrial and Applied Mathematics (SIAM), Philadelphia, PA (1990)
5. Dontchev, A.L., Rockafellar, R.T.: Implicit Functions and Solution Mappings, vol. 543. Springer, 2009
6. Jarník, J., Kurzweil, J.: On conditions on right hand sides of differential relations. Časopis Pro Pěstování Matematiky **102**(4), 334–349 (1977)
7. Rockafellar, R.T.: Proximal subgradients, marginal values, and augmented Lagrangians in nonconvex optimization. Math. Oper. Res. **6**(3), 424–436 (1981)
8. Rockafellar, R.T., Wets, R.J.B.: Variational Analysis, volume 317 of Grundlehren der Mathematischen Wissenschaften [Fundamental Principles of Mathematical Sciences]. Springer, Berlin (1998)
9. Vinter, R.B.: Optimal Control. Birkhäuser, Boston, MA (2000)
10. Vitale, R.A.: The steiner point in infinite dimensions. Israel J. Math. **52**(3), 245–250 (1985)

Chapter 2
Regularity of the Value Function

Abstract This chapter undertakes an examination of the Lipschitz regularity of the value function within the context of infinite horizon control problems, with a particular emphasis on those incorporating discount factors, time-dependent data, and state constraints. To ensure feasibility and to derive local estimates within the ensemble of admissible trajectories, we impose controllability assumptions. We establish the Lipschitz continuity of the value function and demonstrate its asymptotic convergence to zero at infinity within the feasible set—contingent upon a discounted Lagrangian.

Keywords Viability · Neighboring estimates · Lipschitz regularity · Controllability conditions

2.1 The Problem Setting

Consider the following infinite horizon control problem subject to time-dependent state constraints[1]

$$\text{minimize} \quad \int_t^{+\infty} e^{-\lambda s} L(s, x(s), u(s)) ds \tag{$\mathcal{P}$}$$

over all feasible trajectory-control pairs (x, u) satisfying

$$\begin{cases} x'(s) = f(s, x(s), u(s)) & \text{a.e. } s \in \mathbb{R}_t^+, \\ x(t) = \bar{x}, & \\ u(s) \in U(s) & \text{a.e. } s \in \mathbb{R}_t^+, \\ x(s) \in \Omega(s) & \forall s \in \mathbb{R}_t^+, \end{cases} \tag{2.1}$$

[1] We recall that for a function $q \in L^1_{\text{loc}}(\mathbb{R}_t^+; \mathbb{R})$ the integral

$$\int_t^{+\infty} q(s)\, ds := \lim_{T \to +\infty} \int_t^T q(s)\, ds,$$

provided this limit exists.

V. Basco, *Nonsmooth Constrained Optimal Control on Infinite Horizon*, SpringerBriefs in Mathematics, https://doi.org/10.1007/978-3-032-08938-0_2

where $\lambda > 0$ and $s \rightsquigarrow U(s)$, $s \rightsquigarrow \Omega(s)$ are measurable set-valued maps with nonempty closed images.

The optimal control problem described above is applicable to several scenarios within economics and engineering sciences. In these areas of application, the constraint set $\Omega(.)$ manifests as functional constraints

$$\begin{cases} h_1(s, x(s)) \leq 0 \quad \forall s \in \mathbb{R}_t^+, \\ \vdots \\ h_m(s, x(s)) \leq 0 \quad \forall s \in \mathbb{R}_t^+, \end{cases}$$

where h_i's are real valued functions, measurable in time, and space-$\Gamma^{1,\theta}$ regular, uniformly in time. Here $\Gamma^{1,\theta}$ stands for the class of continuously differentiable functions with θ-Höelder continuous and bounded differential, i.e., for $\theta \in]0, 1]$

$$\psi \in \Gamma^{1,\theta} \quad \Longleftrightarrow \quad \begin{cases} \psi \text{ continuously differentiable,} \\ \nabla\psi \text{ bounded, and} \\ \exists k > 0 : |\nabla\psi(x) - \nabla\psi(y)| \leq k|x - y|^{\theta}. \end{cases}$$

A wide class of these functions are represented by affine functions in space with measurable time-dependent terms, specifically, $h_i(s, x) = A(s)x_k + B(s)$ which falls under the framework of the proposed model. This family of functions extends to include $h_i(s, x) = A(s)\psi_i(x) + B(s)$, with $c_i \in \mathbb{R}^n$ a parameter and $\psi_i \in \Gamma^{1,\theta}$.

2.2 Viability

In this section, we develop fundamental results on viability property for set-valued differential inclusions subject to state constraints. Viability ensures the existence of trajectories that remain within an admissible set throughout the considered time horizon. In the next section, we establish sufficient conditions for the existence of such trajectories, under suitable regularity assumptions on the dynamics and constraints.

Lemma 2.2.1 *Let $E : \mathbb{R}_0^+ \rightsquigarrow \mathbb{R}^d$ and $Y : \mathbb{R}_0^+ \rightsquigarrow \mathbb{R}^d$ be set-valued maps with nonempty closed values. Assume that $E(.)$ is continuous of locally bounded variations and $Y(.)$ satisfies:*

(i) $Y(.)$ is measurable for every $x \in \mathbb{R}^d$ with compact images;
(ii) there exists $\rho \in L^1_{\text{loc}}(\mathbb{R}_0^+; \mathbb{R}_0^+)$ such that the Hausdorff distance $d_{\mathscr{H}}(Y(t), Y(s)) \leq \int_s^t \rho(h)dh$ for all $0 \leq s \leq t$.

Let Ψ be the function

$$t \mapsto \Psi(t) := \text{dist}\,(E(t), Y(t)).$$

Then Ψ is continuous and of locally bounded variations on $\mathbb{R}_0^+$.

Proof Notice that, recalling definition (1.8), we can rewrite the Hausdorff distance as $d_{\mathscr{H}}(A, B) = exc(A|B) \vee exc(B|A)$. We first show that the function

$$(t, x) \mapsto d_{E(t)}(x) \text{ is continuous in } \mathbb{R}_0^+ \times \mathbb{R}^d. \tag{2.2}$$

Indeed, from the continuity of E and Lemma 1.3.5 we have that the function $t \mapsto d_{E(t)}(0)$ is continuous on $\mathbb{R}_0^+$. Furthermore, for any $x \in \mathbb{R}^n$, it follows that the map $t \mapsto d_{E(t)}(x)$ is continuous on $\mathbb{R}_0^+$ because, by applying the same argument above, $t \rightsquigarrow E(t)\setminus\{x\}$ is continuous as well. Hence, since for any $t, s \in \mathbb{R}_0^+$ and $x, y \in \mathbb{R}^d$ the triangular inequality yield

$$|d_{E(t)}(x) - d_{E(s)}(y)| \leq |d_{E(t)}(x) - d_{E(s)}(x)| + |x - y|,$$

it follows (2.2).

Now, fix $[a, b] \subset \mathbb{R}_0^+$. Notice that, by the triangular inequality and our assumptions, for every $a \leq s \leq b$

$$\begin{aligned} d_{\mathscr{H}}(\{0\}, Y(s)) &\leq d_{\mathscr{H}}(\{0\}, Y(b)) + d_{\mathscr{H}}(Y(s), Y(b)) \\ &\leq d_{\mathscr{H}}(\{0\}, Y(b)) + \int_a^b \rho(h)dh =: r. \end{aligned}$$

So we deduce that $Y(s) \subset r\mathbb{B}$ for all $s \in [a, b]$. Using (2.2) and the Weierstrass Theorem, we put $R := r + \max\{d_{E(t)}(x) \mid x \in r\mathbb{B}, t \in [a, b]\}$. Hence, it follows for any $s \in [a, b]$

$$\operatorname{dist}(E(s) \cap R\mathbb{B}, Y(s)) = \operatorname{dist}(E(s), Y(s)).$$

Since for any $s, t \in \mathbb{R}_0^+$ and any compact $\mathscr{K} \subset \mathbb{R}^d$

$$E(s) \cap \mathscr{K} \subset E(t) + exc(E(s) \cap \mathscr{K} | E(t))\, \mathbb{B},$$

keeping $\mathscr{K} = R\mathbb{B}$, we get

$$\begin{aligned} &\operatorname{dist}(Y(s), E(s)) \\ &\leq \operatorname{dist}(Y(s), E(t)) + exc(E(s) \cap \mathscr{K} | E(t)) \vee exc(E(t) \cap \mathscr{K} | E(s)) \end{aligned}$$

and

$$\begin{aligned} &\operatorname{dist}(Y(s), E(t)) \\ &\leq \operatorname{dist}(Y(s), E(s)) + exc(E(s) \cap \mathscr{K} | E(t)) \vee exc(E(t) \cap \mathscr{K} | E(s)). \end{aligned}$$

Thus, for every $s, t \in \mathbb{R}_0^+$

$$
\begin{aligned}
&|\text{dist}\,(Y\,(s)\,,E\,(t)) - \text{dist}\,(Y\,(s)\,,E\,(s))| \\
&\le exc\,(E\,(s)\cap\mathscr{K}|E\,(t)) \vee exc\,(E\,(t)\cap\mathscr{K}|E\,(s))\,.
\end{aligned}
$$

Finally, there exists $M > 0$, depending only on $[a, b]$, such that, for any partition $a = t_1 < t_2 < \cdots < t_{m-1} < t_m = b$,

$$
\begin{aligned}
&\sum_{i=1}^{m-1} |\Psi\,(t_{i+1}) - \Psi\,(t_i)| \\
&\le \sum_{i=1}^{m-1} |\text{dist}\,(Y\,(t_{i+1})\,,E\,(t_{i+1})) - \text{dist}\,(Y\,(t_{i+1})\,,E\,(t_i))| \\
&\quad + \sum_{i=1}^{m-1} d_{\mathscr{H}}\,(Y\,(t_{i+1})\,,Y\,(t_i)) \\
&\le \sum_{i=1}^{m-1} exc\,(E\,(t_{i+1})\cap\mathscr{K}|E\,(t_i)) \vee exc\,(E\,(t_i)\cap\mathscr{K}|E\,(t_{i+1})) \\
&\quad + \int_a^b \rho(h)dh \\
&\le M.
\end{aligned}
$$

Hence, the locally bounded variations property for real valued functions follows.

Next we show that Ψ is uniformly continuous in $[a, b]$. Fix $\varepsilon > 0$. For any $\tau \in [a, b]$ consider $y_\varepsilon(\tau) \in Y(\tau)$ such that

$$
\text{dist}(Y(\tau), E(\tau)) + \frac{\varepsilon}{4} \ge d_{E(\tau)}(y_\varepsilon(\tau)).
$$

Then, for any $s, t \in [a, b]$ and any $x_s \in Y(s)$

$$
\begin{aligned}
&\Psi(s) - \Psi(t) \\
&\le \text{dist}(Y(s), E(s)) - d_{E(s)}(y_\varepsilon(t)) + \frac{\varepsilon}{4} \\
&\le d_{E(s)}(x(s)) - d_{E(t)}(y_\varepsilon(t)) + \frac{\varepsilon}{4} \\
&\le |x_s - y_\varepsilon(t)| + |d_{E(s)}(y_\varepsilon(t)) - d_{E(t)}(y_\varepsilon(t))| + \frac{\varepsilon}{4}.
\end{aligned}
\tag{2.3}
$$

We notice that, from our assumptions, there exists $\delta > 0$, depending only on $[a, b]$, such that for every $s, t \in [a, b]$ with $|s - t| \le \delta$

$$
d_{\mathscr{H}}(Y(s), Y(t)) \le \frac{\varepsilon}{4}. \tag{2.4}
$$

Furthermore, applying the triangle inequality and the Lipschitz continuity of the distance function, for any $s, t \in [a, b]$

$$\begin{aligned}|d_{E(s)}(y_\varepsilon(t)) - d_{E(t)}(y_\varepsilon(t))| &\le |d_{E(s)}(y_\varepsilon(s)) - d_{E(s)}(y_\varepsilon(t))| \\ &\quad + |d_{E(s)}(y_\varepsilon(s)) - d_{E(t)}(y_\varepsilon(t))| \\ &\le |y_\varepsilon(s) - y_\varepsilon(t)| + |d_{E(s)}(y_\varepsilon(s)) - d_{E(t)}(y_\varepsilon(t))|.\end{aligned} \tag{2.5}$$

We recall that, from (2.2), the map $(\tau, x) \mapsto d_{E(\tau)}(x)$ is uniformly continuous on $[a, b] \times R\mathbb{B}$, with $R > 0$ depending on $[a, b]$ as above. Hence, replacing δ with a sufficiently small one and using (2.4), for any $s, t \in [a, b]$ with $|s - t| \le \delta$ holds

$$|d_{E(s)}(y_\varepsilon(s)) - d_{E(t)}(y_\varepsilon(t))| \le \frac{\varepsilon}{4}. \tag{2.6}$$

Moreover, from our assumptions and applying the Measurable Selection Theorem, we can find a measurable selection $x(\tau) \in Y(\tau)$ for all $\tau \in \mathbb{R}_0^+$. Thus, keeping $x_s = x(s)$ in (2.3), using (2.4), (2.5), and (2.6), we conclude that

$$\Psi(s) - \Psi(t) \le \varepsilon \quad \forall s, t \in [a, b] \text{ such that } |s - t| \le \delta.$$

From the symmetry with respect to s and t in the previous inequality, the conclusion follows. □

Lemma 2.2.2 *Consider the assumptions of Lemma 2.2.1 with $E(.)$ locally absolutely continuous as set-valued map. Then $\Psi(.)$ is locally absolutely continuous.*

Proof The conclusion follows with the same arguments used in the proof of Lemma 2.2.1 and by definition of locally absolutely continuous set-valued maps given in Sect. 1.3. □

In what follows, we denote by

$$D^+\Psi(t) := \limsup_{h \to 0+} \frac{\Psi(t+h) - \Psi(t)}{h} \in [-\infty, +\infty]$$

the right Dini derivative at $t \in \mathbb{R}$ of a real valued function $\Psi(.)$. Before to state the main result of this section, we need the following version of Gronwall's inequality.

Lemma 2.2.3 *Let $\Psi : [\tau, T] \to \mathbb{R}$ be a continuous function and $\alpha, \beta : [\tau, T] \to \mathbb{R}$ be two locally bounded functions, with $\alpha(.) \ge 0$, such that*

$$D^+\Psi(t) \le \alpha(t)\Psi(t) + \beta(t) \quad \forall t \in]\tau, T[.$$

Then, for every $t \in [\tau, T[$,

$$\Psi(t) \le \Psi(\tau)e^{\alpha(t-\tau)} + \int_\tau^t e^{\alpha(t-r)}\beta\, dr$$

where $\alpha := \sup_{s\in[\tau,T]} \alpha(s)$ *and* $\beta := \sup_{s\in[\tau,T]} |\beta(s)|$.

Proof Let $\delta > 0$ and define $\varphi_\delta(t) := (\Psi(\tau)+\delta)e^{\alpha(t-\tau)} + \int_\tau^t e^{\alpha(t-r)}(\beta+\delta)dr$. Then, $\varphi_\delta'(t) = \alpha\varphi_\delta(t) + \beta + \delta$ in $[\tau, T[$ and $\varphi_\delta(t) > \Psi(\tau)$ for all $t \in [\tau, T[$ close to τ. We show that $\varphi_\delta(.) \geq \Psi(.)$ for any $\delta > 0$. By contradiction, assume that there are some $\bar{t} \in \left]\tau, T\right]$ and $\delta > 0$ with $\varphi_\delta\left(\bar{t}\right) < \Psi\left(\bar{t}\right)$. Setting $s := \inf\left\{t \in \left[\tau, \bar{t}\right] \mid \varphi_\delta(t) < \Psi(t)\right\}$, we obtain that $\varphi_\delta(s) = \Psi(s)$ and $\tau < s < \bar{t}$. Thus, from the definition of s,

$$\begin{aligned}\varphi_\delta'(s) = \liminf_{h\to 0+} \frac{\varphi_\delta(s+h) - \varphi_\delta(s)}{h} &\leq \limsup_{h\to 0+} \frac{\Psi(s+h) - \Psi(s)}{h} \\ &\leq \alpha(s)\Psi(s) + \beta(s) \\ &\leq \alpha\varphi_\delta(s) + \beta,\end{aligned}$$

i.e., $\alpha\varphi_\delta(s) + \beta + \delta \leq \alpha\varphi_\delta(s) + \beta$. Then a contradiction follows. □

Theorem 2.2.4 *Assume that* $E : \mathbb{R}_0^+ \rightsquigarrow \mathbb{R}^d$ *is continuous and of locally bounded variations, let* $t_0 \in \mathbb{R}_0^+$, $x_0 \in E(t_0)$, *and* $\Phi : \mathbb{R}_0^+ \times \mathbb{R}^d \rightsquigarrow \mathbb{R}^d$ *be a set-valued map with nonempty convex closed values such that:*

(i) $\Phi(., x)$ *is measurable for any* $x \in \mathbb{R}^d$;
(ii) $\Phi(t, .)$ *is continuous for any* $t \subset \mathbb{R}_0^+$;
(iii) $\exists\rho \in L^1_{\mathrm{loc}}(\mathbb{R}_0^+; \mathbb{R}_0^+)$: $\sup_{v\in\Phi(t,x)} |v| \leq \rho(t)(1+|x|)$ *for a.e.* $t > 0$.

Then, if for a.e. $t > 0$ *and all* $y \in E(t)$

$$\mathrm{cl\,co}\, T_{\mathrm{graph}\, E}(t, y) \cap (\{1\} \times \Phi(t, y)) \neq \emptyset, \tag{2.7}$$

then for any $t_0 \in \mathbb{R}_0^+$ *and* $x_0 \in E(t_0)$ *there exists a locally absolutely continuous viable solution*

$$\begin{cases} x'(t) \in \Phi(t, x(t)) & a.e.\ t > t_0, \\ x(t_0) = x_0, \\ x(t) \in E(t) & \forall t > t_0. \end{cases}$$

We need first the following lemma.

Lemma 2.2.5 *Let* I *be a compact interval of the real line and consider* $\Phi : I \times \mathbb{R}^d \rightsquigarrow \mathbb{R}^d$ *a set-valued map with nonempty closed convex values satisfying:* $\Phi(t, .)$ *is continuous for a.e.* t; $\Phi(., x)$ *is measurable for all* x; $\sup_{\phi\in\Phi(t,x)} |\phi| \leq q(t)(1+|x|)$ *for all* $x \in \mathbb{R}^d$ *and a suitable* $q \in L^1(I; \mathbb{R}_0^+)$. *Then there exists a set of full measure* $\mathscr{B} \subset I$ *such that:*

- *any* $(\tau, x_\tau) \in \mathscr{B} \times \mathbb{R}^d$ *and any* $\varepsilon > 0$ *and there exists* $\eta > 0$ *such that for all* $x \in \mathrm{Ker}\,(\Phi, \tau, x_\tau)$ *and all* $h \in]-\eta, +\eta[\setminus\{0\}$

$$\frac{x(\tau+h) - x_\tau}{h} \in \Phi(\tau, x_\tau) + \varepsilon\mathbb{B};$$

– *for any* $(\tau, x_\tau) \in \mathscr{B} \times \mathbb{R}^d$ *and any* $\overline{\phi} \in \Phi(\tau, x_\tau)$ *there exists a solution of the problem* $x'(t) \in \Phi(t, x(t))$ *for a.e.* t *in a neighborhood of* τ, $x(\tau) = x_\tau$, *and* $x'(\tau) = \overline{\phi}$.

In particular,

$$\emptyset \neq \mathrm{Limsup}_{h \to 0+} \left\{ \frac{x(\tau + h) - x_\tau}{h} \right\} \subset \Phi(\tau, x_\tau),$$
$$\emptyset \neq \mathrm{Limsup}_{h \to 0+} \left\{ \frac{x(\tau - h) - x_\tau}{-h} \right\} \subset \Phi(\tau, x_\tau).$$

Proof See Sect. A.3.1. □

Proposition 2.2.6 *All the conclusions of Theorem 2.2.4 hold true if the continuity condition in Theorem 2.2.4:(ii) is replaced by*

$$\forall r > 0, \ \exists k_r \in L^1_{\mathrm{loc}}(\mathbb{R}^+_0; \mathbb{R}^+_0): \quad \Phi(t, .) \text{ is } k_r(t) - \textit{Lipschitz on } r\mathbb{B}, \textit{ for a.e. } t. \tag{2.8}$$

Proof First of all notice that by Gronwall's Lemma and our sub-linear growth assumption, for all $r > 0$ there exists $R > r$ such that if an absolutely continuous function $x : [0, t_1] \to \mathbb{R}^d$ satisfies $|x'(t)| \leq \rho(t)(1 + |x(t)|)$ a.e. in $[0, t_1]$, $x(0) = x_0$, and $|x_0| \leq r$, then $|x(t)| \leq R$ for all $t \in [0, t_1]$. Moreover, observe that Φ is integrably bounded on $[0, t_1]$, i.e. for almost all $t \in [0, t_1]$ and all $x \in r\mathbb{B}$,

$$|v| \leq \rho(t)(1 + R) := \rho_R(t) \tag{2.9}$$

for any $v \in \Phi(t, x)$.

Now, fix $t_0 \in \mathbb{R}^+_0$, $x_0 \in E(t_0)$, and let any $T > t_0$. Consider the map Ψ of the Lemma 2.2.1 applied with $Y(.)$ defined by the reachable set

$$Y(s) = R[t_0, x_0](s) := \{x(s) \mid x \in \mathrm{Ker}(\Phi, t_0, x_0)\}.$$

From Lemmata 2.2.1 and 2.2.5, there exists a subset $\mathscr{C} \subset [0, T]$ of full measure such that for all $t \in \mathscr{C}$ and $x \in \mathbb{R}^d$ the following properties hold:

$$\begin{cases} \Psi \text{ is differentiable at } t; \\ \forall v \in \Phi(t, x) \text{ there exists } y \in \mathrm{Ker}(\Phi, t, x) \text{ on } [0, T] \text{ such that} \\ \quad y'(t) = v; \\ \forall s < t, \forall y \in \mathrm{Ker}(\Phi, t, x) \text{ on } [s, t] \text{ and } \forall h_i \to 0+, \\ \quad \emptyset \neq \mathrm{Limsup}_{i \to +\infty}\{\frac{y(t-h_i)-x}{h_i}\} \subset -\Phi(t, x). \end{cases}$$

We claim that

$$\Psi(t) = 0 \quad \forall t > t_0, \tag{2.10}$$

arguing by contradiction. Suppose that $\Psi(T) > 0$ and consider

$$\tau = \sup\{t < T \mid \Psi(t) = 0\}.$$

So, $\Psi > 0$ on $]\tau, T]$ and $\Psi(\tau) = 0$. We divide the proof of the claim (2.10) in two steps.

STEP 1: Let $t \in \mathscr{C}$, $z \in Y(t)$, $y \in E(t)$ satisfying $\Psi(t) = |z - y|$ and put $p = \frac{z-y}{|z-y|}$. We first prove that for all $(u, w) \in T_{\text{graph } E}(t, y)$ the set defined by

$$\Lambda(u, w) := \{v \in \Phi(t, z) \mid \Psi'(t)u \leq \langle p, uv - w\rangle\}$$

satisfy

$$\Lambda(u, w) = \begin{cases} \Phi(t, z) & \text{if } u \geq 0, \\ \emptyset & \text{if } u < 0. \end{cases} \tag{2.11}$$

Indeed, let $(u, w) \in T_{\text{graph } E}(t, y)$ and $h_i \to 0+$, $u_i \to u$, $w_i \to w$ satisfying $y + h_i w_i \in E(t + h_i u_i)$ for all $i \in \mathbb{N}$. Suppose that there exists a subsequence $(u_{i_k})_{k\in\mathbb{N}}$ with $u_{i_k} \geq 0$ for all $k \in \mathbb{N}$. Let $v \in \Phi(t, z)$ and $x(.)$ a solution of (2.18) on $[t, T]$ such that $x(t) = z$ and $x'(t) = v$. Thus

$$\Psi\left(t + h_{i_k}u_{i_k}\right) - \Psi(t) \leq |x\left(t + h_{i_k}u_{i_k}\right) - y - h_{i_k}w_{i_k}| - |z - y|.$$

Dividing by h_{i_k} and taking the limit we get $\Psi'(t)u \leq \langle p, uv - w\rangle$. Otherwise, we have $u_i < 0$ for all i large enough. In this case, consider a solution $\bar{x}(.)$ of (2.18) on $[t_0, t]$, $(i_k)_{k\in\mathbb{N}}$, and $\bar{v} \in \Phi(t, z)$ such that $\bar{x}(t_0) = x_0$, $\bar{x}(t) = z$, and $\lim_{k\to+\infty} \frac{\bar{x}(t+h_{i_k}u_{i_k})-z}{h_{i_k}} = u\bar{v}$. Hence for all $k \in \mathbb{N}$ sufficiently large

$$\Psi\left(t + h_{i_k}u_{i_k}\right) - \Psi(t) \leq |\bar{x}\left(t + h_{i_k}u_{i_k}\right) - y - h_{i_k}w_{i_k}| - |z - y|.$$

Dividing by h_{i_k} and taking the limit we get $\Psi'(t)u \leq \langle p, u\bar{v} - w\rangle$. Hence, it follows (2.11). Now, consider $e_j \geq 0$ and $\left(u_j, w_j\right) \in T_{\text{graph } E}(t, y)$ for $j = 0, \ldots, d$ such that $\sum_{j=0}^{d} e_j = 1$ and $u := \sum_{j=0}^{d} e_j u_j > 0$. Without loss of generality, we may assume that for some natural number $0 \leq N < d$ and all $j = 1, \ldots, N$ we have $u_j \geq 0$ and $u_j < 0$ for all $j = N+1, \ldots, d$. From (2.11), for every $j = N+1, \ldots, d$ there exists $\bar{v}_j \in \Phi(t, z)$ such that $\Psi'(t)u_j \leq \left\langle p, u_j\bar{v}_j - w_j\right\rangle$. Thus, applying again (2.11) it follows that

$$\begin{aligned} \Psi'(t)(\sum_{j=0}^{N} e_j u_j) &\leq \langle p, \sum_{j=0}^{N} e_j u_j v - \sum_{j=0}^{N} e_j w_j\rangle \quad \forall v \in \Phi(t, z), \\ \Psi'(t)(\sum_{j=N+1}^{d} e_j u_j) &\leq \langle p, \sum_{j=N+1}^{d} e_j u_j \bar{v}_j - \sum_{j=N+1}^{d} e_j w_j\rangle. \end{aligned} \tag{2.12}$$

Notice that, since $e_{N+1}|u_{N+1}| + \cdots + e_d|u_d| = |e_{N+1}u_{N+1} + \cdots + e_d u_d| < e_0u_0 + \cdots + e_N u_N$, we have

$$\theta_j := \frac{e_j |u_j|}{\sum_{j=0}^{N} e_j u_j} \implies \sum_{j=N+1}^{d} \theta_j \in]0, 1[,$$

that, due to convexity of $\Phi(t, z)$, it implies in turn that $(1 - \sum_{j=N+1}^{d} \theta_j)v + \sum_{j=N+1}^{d} \theta_j \bar{v}_j \in \Phi(t, z)$ for every $v \in \Phi(t, z)$. Hence, recalling (2.11) and (2.12), we obtain for all $v \in \Phi(t, z)$

$$\begin{aligned}\Psi'(t)u = \Psi'(t)\Big(\sum_{j=0}^{d} e_j u_j\Big) &\le \langle p, (\sum_{j=0}^{N} e_j u_j - \sum_{j=N+1}^{d} e_j |u_j|)v \\ &\quad + \sum_{j=N+1}^{d} e_j(|u_j| + u_j)\bar{v}_j - \sum_{j=0}^{d} e_j w_j\rangle \\ &= \langle p, uv - \sum_{j=0}^{d} e_j w_j\rangle.\end{aligned}$$

So, we have that:

$$\begin{cases} \forall t \in \mathscr{C}, \forall (u, w) \in \text{cl co } T_{\text{graph } E}(t, y), \text{ with } u > 0, \\ \exists h_i \to 0+, \exists u_i \to u, \exists w_i \to w : \\ y + h_i w_i \in E(t + h_i u_i) \text{ for all } i \in \mathbb{N} \text{ and} \\ \lim_{i \to +\infty} \frac{\Psi(t+h_i u_i) - \Psi(t)}{h_i} = \Psi'(t)u \le \langle p, uv - w\rangle, \ \forall v \in \Phi(t, z), \end{cases}$$

where $z \in Y(t)$, $y \in E(t)$ satisfy $\Psi(t) = |z - y|$, and $p := \frac{z-y}{|z-y|}$.

STEP 2: Fix $\varepsilon > 0$ and $t \in]\tau, T[$. Keep a sequence of positive numbers $h_k \to 0+$ such that $\lim_{k\to+\infty} \frac{\Psi(t+h_k)-\Psi(t)}{h_k} = D^+\Psi(t)$. From Lemma 2.2.1, for all $k \in \mathbb{N}$ there exists $\delta_k > 0$ such that for any sequences $(s_k)_{k\in\mathbb{N}}$, $(\tilde{s}_k)_{k\in\mathbb{N}} \subset [\tau, T]$

$$(|s_k - \tilde{s}_k| \le \delta_k \quad \forall k \in \mathbb{N}) \implies (|\Psi(s_k) - \Psi(\tilde{s}_k)| \le o(h_k) \quad \forall k \in \mathbb{N}). \tag{2.13}$$

Moreover, applying again Lemma 2.2.1, we can find a sequence of differentiability points $(t_k)_k \subset \mathscr{C}$ for Ψ such that

$$\varepsilon h_k + \Psi(t) \ge \Psi(t_k), \ |t_k - (t + h_k)| \le \frac{\delta_k}{3} \quad \forall k \in \mathbb{N}. \tag{2.14}$$

For any $k \in \mathbb{N}$ consider $z_k \in Y(t_k)$ and $y_k \in E(t_k)$ such that $\Psi(t_k) = |z_k - y_k|$ and put $p_k = \frac{z_k - y_k}{|z_k - y_k|}$. From Step 1, for any $k \in \mathbb{N}$ and any $(u_k, w_k) \in \text{cl co } T_{\text{graph } E}(t_k, y_k)$, with $u_k > 0$, there exist $h_j^{(k)} \to 0+$, $u_j^{(k)} \to u_k$, and $w_j^{(k)} \to w_k$ satisfying

$$y_k + h_j^{(k)} w_j^{(k)} \in E(t_k + h_j^{(k)} u_j^{(k)}) \quad \forall j \in \mathbb{N},$$

and

$$\lim_{j \to +\infty} \frac{\Psi(t_k + h_j^{(k)} u_j^{(k)}) - \Psi(t_k)}{h_j^{(k)}} \leq \langle\, p_k, u_k v - w_k \rangle \quad \forall v \in \Phi(t_k, z_k).$$

In particular, it follows that for any $k \in \mathbb{N}$ we can choose $j_k \in \mathbb{N}$ satisfying

$$|h_j^{(k)}| \leq h_k, \quad |h_j^{(k)} u_j^{(k)}| \leq \frac{\delta_k \wedge h_k}{3} \quad \forall j \geq j_k. \tag{2.15}$$

Hence, from (2.13) and (2.14), we have for any large $k \in \mathbb{N}$ and any $j \geq j_k$

$$\frac{\Psi(t + h_k) - \Psi(t_k + h_j^{(k)} u_j^{(k)})}{h_k} = o(h_k)$$

and

$$\begin{aligned} &\frac{\Psi(t + h_k) - \Psi(t)}{h_k} \\ &\leq \frac{\Psi(t + h_k) - \Psi(t_k + h_j^{(k)} u_j^{(k)})}{h_k} + \frac{\Psi(t_k + h_j^{(k)} u_j^{(k)}) - \Psi(t_k)}{h_k} + \varepsilon. \end{aligned} \tag{2.16}$$

From (2.15) and Step 1, we get for every large $k \in \mathbb{N}$

$$\limsup_{j \to +\infty} \frac{\Psi(t_k + h_j^{(k)} u_j^{(k)}) - \Psi(t_k)}{h_k} \leq \langle\, p_k, u_k v - w_k \rangle \quad \forall v \in \Phi(t_k, z_k).$$

Using that, inequality in (2.14), assumption (2.8), and since for all $k \in \mathbb{N}$ we can find $(u_k, w_k) = (1, v_k) \in \text{cl co}\, T_{\text{graph}\, E}(t_k, y_k)$ with $v_k \in \Phi(t_k, y_k)$, passing in (2.16) to the upper limit as $j \to +\infty$ we get for any $v_z \in \Phi(t_k, z_k)$

$$\begin{aligned} \frac{\Psi(t + h_k) - \Psi(t)}{h_k} &\leq o(h_k) + \langle\, p_k, v_z - v_k \rangle + \varepsilon \\ &\leq o(h_k) + k_r(t_k)|z_k - y_k| + \varepsilon \\ &= o(h_k) + k_r(t_k)\Psi(t_k) + \varepsilon \\ &\leq o(h_k)(1 + k_r \varepsilon) + \varepsilon + k_r \Psi(t) \quad \text{for all large } k \in \mathbb{N}, \end{aligned} \tag{2.17}$$

where $k_r = \sup_{t \in [\tau, T]} |k_r(t)|$. Passing first to the limit in (2.17) as $k \to +\infty$, and then using the arbitrariness of ε, we get

$$D^+ \Psi(t) \leq k_r \Psi(t).$$

From that and Lemma 2.2.3, the claim (2.10) follows.

We show now that for any $T > t_0$ and any $t_0 \leq t_i < t_{i+1} \leq T, i = 1, \ldots, m$, there exists on $[t_0, T]$ a solution $x(.)$ of

$$x'(t) \in \Phi(t, x(t)) \text{ a.e. } t \tag{2.18}$$

satisfying $x(t_0) = x_0$ and

$$x(t_i) \in E(t_i) \quad \forall i = 1, \ldots, m+1. \tag{2.19}$$

Consider $t_0 \leq t_i < t_{i+1} \leq T$, $i = 1, \ldots, m$. Following the induction argument, assume that for some $j \geq 0$ there exists an absolutely continuous trajectory $y : [t_0, t_i] \to \mathbb{R}^d$ solving (2.18) such that $y(t_i) \in E(t_i)$ for all $i \leq j$. From the above claim applied with (t_0, x_0) replaced by $(t_j, y(t_j))$ we can find an absolutely continuous trajectory $\bar{y} : [t_j, T] \to \mathbb{R}^d$ solving (2.18) such that $\bar{y}(t_j) = y(t_j)$ and $\bar{y}(t_{j+1}) \in E(t_{j+1})$. Thus we can extend y on the time interval $[t_j, t_{j+1}]$ by setting $y(t) = \bar{y}(t)$ for all $t \in [t_j, t_{j+1}]$.

To conclude the proof, let $t_0 \in \mathbb{R}_0^+$, $x_0 \in E(t_0)$, $T > t_0$, and $\varepsilon > 0$. Pick $\delta > 0$ such that the map Ψ in the Lemma 2.2.1 is uniformly continuous in $[t_0, T]$. By replacing δ with a suitable small one, we can assume that $\int_s^{\bar{s}} \rho(t)dt \leq \varepsilon$ for any $s, \bar{s} \in [t_0, T]$ such that $|s - \bar{s}| \leq \delta$. Let a finite partition $t_0 < t_1 < \cdots < t_m = T$ be such that $t_{i+1} - t_i < \delta$. Then, from the choice of δ, Gronwall's Lemma, and (2.19), there exist a constant $c > 0$ (depending only on $|x_0|$ and T) and a trajectory $x_\varepsilon(.)$ solving (2.18) on $[t_0, T]$ such that for any $t \in [t_0, T]$ and i with $t_i \leq t \leq t_{i+1}$

$$\begin{aligned}
d_{E(t)}(x_\varepsilon(t)) &\leq \operatorname{dist}(Y_i(t), E(t)) + d_{Y_i(t)}(x_\varepsilon(t)) \\
&= \operatorname{dist}(Y_i(t), E(t)) - \operatorname{dist}(Y_i(t_{i+1}), E(t_{i+1})) + d_{Y_i(t)}(x_\varepsilon(t)) \\
&\leq \varepsilon + c \int_{t_i}^{t} \rho(s)ds \\
&\leq (1 + c)\varepsilon,
\end{aligned}$$

where $Y_i(s) = R[t_i, x(t_i)](s)$. Now, applying again Gronwall's Lemma, consider a sequence $(x_{\varepsilon_j}(.))_{j \in \mathbb{N}}$ converging weakly in $W^{1,1}([t_0, T]; \mathbb{R}^d)$ to some absolutely continuous function $x : [t_0, T] \to \mathbb{R}^d$, where $\varepsilon_j \to 0+$. It follows that x solves (2.18), with $x(t_0) = x_0$, and $x(t) \in E(t)$ for all $t \in [t_0, T]$. Using an iterative argument, we can extend such solution to the whole halfline $\mathbb{R}_{t_0}^+$ in order to get the statement. □

Proof of Theorem 2.2.4 From Lemma 1.3.1 we can find a sequence of increasing closed sets

$$\tilde{W}_k \subset \tilde{W}_{k+1} \subset [t_0, T] \quad \forall k \in \mathbb{N}$$

such that $\mu_{\mathscr{L}}([t_0, T] \setminus \tilde{W}_k) < 1/k$ and $\Phi(.,.)$ is upper semicontinuous on $\tilde{W}_k \times \mathbb{R}^d$. Now, we claim: there exist a sequence of closed sets $(W_k)_{k \in \mathbb{N}} \subset [t_0, T]$ such that $\mu_{\mathscr{L}}([t_0, T] \setminus \cup_k W_k) = 0$, $\Phi(.,.)$ is upper semicontinuous on $W_k \times \mathbb{R}^d$ for all $k \in \mathbb{N}$, and

$$\left(\varphi : s \mapsto \sum_k \sup_{s \in W_k} |\rho_R(s)| \chi_{W_k \setminus W_{k-1}}(s)\right) \in L^1([t_0, T]; \mathbb{R}_0^+)$$

where $\rho_R(.)$ is defined in (2.9). Indeed, consider a sequence of decreasing open sets $(O_k)_k$

$$O_{k+1} \subset O_k \subset [t_0, T] \quad \forall k \in \mathbb{N}$$

such that

$$H_k := \{s \in [t_0, T] \mid \rho_R(s) \geq k\} \subset O_k \quad \forall k \in \mathbb{N},$$

$$\mu_{\mathcal{L}}(O_k \setminus H_k) < \frac{1}{k+1} \quad \forall k \in \mathbb{N}.$$

Let for any $k \in \mathbb{N}$,

$$W_k := \tilde{W}_k \setminus O_k.$$

Of course $W_k \subset W_{k+1}$ for all $k \in \mathbb{N}$ and for any $N \in \mathbb{N}$

$$\begin{aligned}
\mu_{\mathcal{L}}([t_0, T] \setminus \cup_{k=1}^{N} W_k) &= \mu_{\mathcal{L}}([t_0, T] \setminus W_N) \\
&= \mu_{\mathcal{L}}([t_0, T] \setminus (\tilde{W}_N \setminus O_N)) \\
&\leq \mu_{\mathcal{L}}(O_N) \\
&\leq \mu_{\mathcal{L}}(O_N \setminus H_N) + \mu_{\mathcal{L}}(H_N).
\end{aligned} \tag{2.20}$$

Since $\rho_R(.)$ is integrable, we have $\mu_{\mathcal{L}}(H_N) \to 0$ as $N \to +\infty$. Then, passing to the limit in (2.20) we conclude that

$$\mu_{\mathcal{L}}([t_0, T] \setminus \cup_{k=1}^{+\infty} W_k) = \lim_{N \to +\infty} \mu_{\mathcal{L}}([t_0, T] \setminus \cup_{k=1}^{N} W_k) = 0.$$

Furthermore, $\Phi(.,.)$ is upper semicontinuous on $W_k \times \mathbb{R}^d$ by construction of W_k. Furthermore, recalling that $W_k \cap H_k = \emptyset$ for all $k \in \mathbb{N}$, we have

$$\begin{aligned}
\int_{[t_0,T]} \varphi(s) ds &\leq \sum_{k=1}^{+\infty} k \mu_{\mathcal{L}}(W_k \setminus W_{k-1}) \\
&= T - t_0 + \sum_{k=1}^{+\infty} (k+1) \mu_{\mathcal{L}}(W_{k+1} \setminus W_k) \\
&\leq T - t_0 + \sum_{k=1}^{+\infty} \mu_{\mathcal{L}}(W_{k+1} \setminus W_k) + \sum_{k=1}^{+\infty} k \mu_{\mathcal{L}}(W_{k+1} \setminus W_k) \\
&\leq 2(T - t_0) + \int_{[t_0,T]} \rho_R(s) ds,
\end{aligned}$$

and the claim is then proved.

Now, applying Lemma A.2.1 (see Appendix A), for any $k \in \mathbb{N}$ there exists a set-valued map $\Phi_k(.,.)$, with convex compact nonempty values such that: $\mu_{\mathscr{L}}([t_0, T] \setminus \cup_k W_k) = 0$ and for all $k \in \mathbb{N}$,

$$\widehat{\Phi}_{k+1}(t, x) \subset \widehat{\Phi}_k(t, x) \subset \text{cl co}\, \Phi(W_k \times \mathbb{R}^d), \quad \forall t \in W_k, \forall x \in \mathbb{R}^d, \forall k \in \mathbb{N},$$

$$\Phi(t, x) = \cap_k \widehat{\Phi}_k(t, x), \quad \forall t \in \cup_k W_k, \forall x \in \mathbb{R}^d,$$

$$\widehat{\Phi}_k(t, .) \text{ is locally Lipschitz } \forall t \in W_k, \forall k \in \mathbb{N}.$$

Notice that $\left\| \Phi(W_k \times \mathbb{R}^d) \right\| \leq \sup_{s \in W_k} |\rho_R(s)| < +\infty$ for all $k \in \mathbb{N}$. Let us denote for any $k \in \mathbb{N}$ the set-valued map

$$\Phi_k(t, x) := \begin{cases} \widehat{\Phi}_k(t, x) & t \in W_k, x \in \mathbb{R}^d \\ \varphi(t)\mathbb{B} & t \notin W_k, x \in \mathbb{R}^d. \end{cases}$$

The set-valued maps Φ_k satisfies all assumptions of Theorem 2.2.4. Then for any $k \in \mathbb{N}$ there exists $x_k(.) \in \text{Ker}(\Phi_k, \bar{x}, \bar{t})$ such that $x_k(t) \in E(t)$ for all $T > t > t_0$. Denoting such sets with $\mathscr{G}_k$, we have that they are nonempty and compact in $C([t_0, T]; \mathbb{R}^d)$ and $\mathscr{G}_{k+1} \subset \mathscr{G}_k$. Hence, $\cap_k \mathscr{G}_k \neq \emptyset$ and

$$\cap_k \mathscr{G}_k = \{x(.) \in \text{Ker}(\Phi, \bar{x}, \bar{t}) \mid x(t) \in E(t)\, \forall t \in [t_0, T]\}.$$

The proof is now complete. □

2.3 Neighboring Feasible Trajectory Estimates

This section investigates estimates for neighboring feasible trajectories that will serve to establish regularity properties of the value function. We provide uniform linear estimates for state-constrained differential inclusions

$$\begin{cases} x'(t) \in F(t, x(t)) & \text{a.e. } t, \\ x(t) \in \Omega(t) & \forall t, \end{cases}$$

where $(t, x) \rightsquigarrow F(t, x)$ is a given set-valued map taking values in $\mathbb{R}^n$. A function $t \mapsto x(t)$ is said to be:

- *feasible* if $x(t) \in \Omega(t)$ for all t;
- *F-trajectory* if it is absolutely continuous and $x'(t) \in F(t, x(t))$ for a.e. t;
- *feasible F-trajectory* if $x(.)$ is an F-trajectory and it is feasible.

Whenever $x(.)$ is a feasible F-trajectory with $F(t,x) = \{f(t,x,u) \mid u \in U(t)\}$ and starting from $x(t_0) = x_0$, any measurable selection $t \mapsto u(t)$ such that $x'(t) = f(t, x(t), u(t))$ a.e. t is said *feasible at* (t_0, x_0). A function $x : \mathbb{R}^+_{t_0} \to \mathbb{R}^n$ is said to be:

- F_∞*-trajectory* if $x|_{[t_0,t_1]}(.)$ is an *F-trajectory* for all $t_1 > t_0$;
- *feasible* F_∞*-trajectory* if $x|_{[t_0,t_1]}(.)$ is a *feasible F-trajectory* for all $t_1 > t_0$.

Assumption 2.3.1 We assume the following on $F(.,.)$:

(I) $F(.,.)$ has closed and nonempty values, a sub-linear growth, and $F(.,x)$ is Lebesgue measurable for all $x \in \mathbb{R}^n$;
(II) there exist $M \geq 0$ and $\alpha > 0$ such that

$$\sup\{|v| \mid v \in F(t,x),\ t \in \mathbb{R}^+_0,\ x \in \mathrm{bdr}\,\Omega(t) + \alpha\mathbb{B}\} \leq M;$$

(III) there exists $\varphi \in \mathcal{L}_{\mathrm{loc}}$ such that $F(t,.)$ is $\varphi(t)$-Lipschitz continuous for all $t \in \mathbb{R}^+_0$;
(IV) there exist $\tilde{\eta} > 0$ and $\gamma \in \mathcal{L}_{\mathrm{loc}}$ such that F is γ-left absolutely continuous, uniformly wrt $\mathrm{bdr}\,\Omega + \tilde{\eta}\mathbb{B}$.

Assumption 2.3.2 There exist $\varepsilon > 0$ and $\eta > 0$ such that:

$$\begin{cases} \forall t \in \mathbb{R}^+_0,\ \forall x \in (\mathrm{bdr}\,\Omega(t) + \eta\mathbb{B}) \cap \Omega(t), \\ \exists v \in \mathrm{co}\,F(t,x) \text{ satisfying} \\ \{y + [0,\varepsilon](v + \varepsilon\mathbb{B}) \mid y \in (x + \varepsilon\mathbb{B}) \cap \Omega(t)\} \subset \Omega(t). \end{cases}$$

We recall that, in the literature, a map $E : \mathbb{R}^+_0 \rightsquigarrow \mathbb{R}^d$ is also called *tube*.

Theorem 2.3.3 (Neighboring Feasible Estimates) *Consider Assumption 2.3.1, 2.3.2. If there exists $L \geq 0$ such that*

$$\text{the set-valued map } \Omega : \mathbb{R}^+_0 \rightsquigarrow \mathbb{R}^n \text{ is } L-\text{Lipschitz continuous,} \tag{2.21}$$

then for every $\delta > 0$ there exists a constant $\beta > 0$ such that for any $[t_0, t_1] \subset \mathbb{R}^+_0$ with $t_1 - t_0 = \delta$, any F-trajectory $\hat{x}(.)$ defined on $[t_0, t_1]$ with $\hat{x}(t_0) \in \Omega(t_0)$, and any $\varrho > 0$ satisfying

$$\sup_{t \in [t_0,t_1]} d_{\Omega(t)}(\hat{x}(t)) \leq \varrho$$

we can find an F-trajectory $x(.)$ on $[t_0, t_1]$ such that

$$\begin{cases} x(t_0) = \hat{x}(t_0), \\ \|\hat{x} - x\|_{\infty,[t_0,t_1]} \leq \beta\varrho, \\ x(t) \in \mathrm{int}\,\Omega(t) \quad \forall t \in]t_0, t_1]. \end{cases}$$

Proof Fix $\delta > 0$ and let $[t_0, t_1] \subset \mathbb{R}_0^+$ with $t_1 - t_0 = \delta$.

We first show the statement whenever $F = \text{co}\, F$. Let

$$\varepsilon > 0,\ k > 0,\ \Delta > 0,\ \bar{\varrho} > 0, \text{ and } m \in \mathbb{N}^+ \tag{2.22}$$

be such that

$$\Delta \leq \varepsilon, \quad \bar{\varrho} + M\Delta < \varepsilon, \quad k\bar{\varrho} < \varepsilon, \quad k > 1/\varepsilon, \quad 4\Delta M \leq \hat{\eta}, \tag{2.23}$$

$$(i)\ e^{\theta_\varphi(\Delta)}(\theta_\gamma(\Delta) + \theta_\varphi(\Delta)M) < \varepsilon, \ (ii)\ 2e^{\theta_\varphi(\Delta)}(\theta_\gamma(\Delta) + \theta_\varphi(\Delta)M)k < (k\varepsilon - 1), \tag{2.24}$$

and

$$\frac{\delta}{m} \leq \Delta.$$

Notice that all the constants appearing in (2.22) do not depend on the time interval $[t_0, t_1]$, the trajectory $\hat{x}(.)$, and ϱ.

Case 1: $\varrho \leq \bar{\varrho}$ and $\delta \leq \Delta$.

We observe that, by the last inequality in (2.23), if

$$\hat{x}(t_0) \in \Omega(t_0) \backslash (\text{bdr}\ \Omega(t_0) + \frac{\hat{\eta}}{2}\mathbb{B})$$

then $x(.) = \hat{x}(.)$ is as desired. Indeed, without loss of generality, assume $\hat{x}(t_0) \in (\text{bdr}\ \Omega(t_0) + \frac{\hat{\eta}}{2}\mathbb{S}) \cap \Omega(t_0)$ and suppose by contradiction that

$$\mathcal{R} := \{t \in]t_0, t_1] \mid \hat{x}(t) \in \text{cl}\ (\mathbb{R}^n \backslash \Omega(t))\} \neq \emptyset.$$

Put $s := \inf \mathcal{R}$ and notice that $s \neq t_0$. Then, we have

$$d_{\text{bdr}\ \Omega(t_0)}(\hat{x}(s)) \leq \text{dist}(\text{bdr}\ \Omega(t_0), \text{bdr}\ \Omega(s)) + d_{\text{bdr}\ \Omega(s)}(\hat{x}(s)) \leq L|s - t_0|.$$

Since

$$d_{\text{bdr}\ \Omega(t_0)}(\hat{x}(t_0)) - d_{\text{bdr}\ \Omega(t_0)}(\hat{x}(s)) \leq |\hat{x}(s) - \hat{x}(t_0)|,$$

it follows that

$$\hat{\eta}/4 - L|s - t_0| \leq M\Delta \leq \hat{\eta}/4,$$

a contradiction. Next we assume that $\hat{x}(t_0) \in (\text{bdr}\ \Omega(t_0) + \frac{\hat{\eta}}{2}\mathbb{B}) \cap \Omega(t_0)$.

Let $v \in F(t_0, \hat{x}(t_0))$ be as in Assumption 2.3.2 and define $y : [t_0, t_1] \to \mathbb{R}^n$ by

$$y(t_0) := \hat{x}(t_0), \quad y'(t) := \begin{cases} v & t \in [t_0, (t_0 + k\varrho) \wedge t_1], \\ \hat{x}'(t - k\varrho) & t \in]t_0 + k\varrho, t_1] \cap J, \end{cases} \tag{2.25}$$

where $J = \{s \in]t_0 + k\varrho, t_1] \,|\, \hat{x}'(s - k\varrho) \text{ exists}\}$. Hence

$$\|\hat{x} - y\|_{\infty,[t_0,t_1]} \leq 2Mk\varrho. \tag{2.26}$$

By the Generalized Filippov Existence Theorem there exists an F-trajectory $x(.)$ on $[t_0, t_1]$ such that $x(t_0) = y(t_0)$ and

$$\|y - x\|_{\infty,[t_0,t]} \leq e^{\int_{t_0}^{t} \varphi(\tau)\,d\tau} \int_{t_0}^{t} d_{F(s,y(s))}(y'(s))\,ds \tag{2.27}$$

for all $t \in [t_0, t_1]$. Then, using Assumption 2.3.1:*(III)*, (1.9), and (2.25), it follows that

$$d_{F(s,y(s))}(y'(s)) \leq \begin{cases} \theta_\gamma(\Delta) + \varphi(s)M(s - t_0) & \text{a.e. } s \in [t_0, (t_0 + k\varrho) \wedge t_1], \\ \varphi(s)Mk\varrho + \int_{s-k\varrho}^{s} \gamma(\tau)\,d\tau & \text{a.e. } s \in]t_0 + k\varrho, t_1]. \end{cases}$$

Hence, we obtain for any $t \in [t_0, (t_0 + k\varrho) \wedge t_1]$

$$\int_{t_0}^{t} d_{F(s,y(s))}(y'(s))\,ds \leq (\theta_\gamma(\Delta) + \theta_\varphi(\Delta)M)(t - t_0),$$

and, using the Fubini Theorem, for any $t \in]t_0 + k\varrho, t_1]$

$$\int_{t_0+k\varrho}^{t} d_{F(s,y(s))}(y'(s))\,ds \leq (\theta_\varphi(\Delta)M + \theta_\gamma(\Delta))k\varrho.$$

Thus, by (2.27), for all $t \in [t_0, (t_0 + k\varrho) \wedge t_1]$

$$\|y - x\|_{\infty,[t_0,t]} \leq e^{\theta_\varphi(\Delta)}(\theta_\gamma(\Delta) + \theta_\varphi(\Delta)M)(t - t_0), \tag{2.28}$$

and

$$\|y - x\|_{\infty,[t_0,t_1]} \leq 2e^{\theta_\varphi(\Delta)}(\theta_\gamma(\Delta) + \theta_\varphi(\Delta)M)k\varrho. \tag{2.29}$$

Finally, taking note of (2.26), it follows that

$$\|\hat{x} - x\|_{\infty,[t_0,t_1]} \leq \beta_1\varrho,$$

where we put $\beta_1 = 2(M + e^{\theta_\varphi(\Delta)}(\theta_\gamma(\Delta) + \theta_\varphi(\Delta)M))k$.

We claim next that

$$x(t) \in \operatorname{int}\Omega(t) \quad \forall t \in]t_0, t_1].$$

Indeed, if $t \in]t_0, (t_0 + k\varrho) \wedge t_1]$, then from Assumption 2.3.2, the first condition in (2.23), and (2.25) it follows that

$$y(t) + (t - t_0)\varepsilon\mathbb{B} = \hat{x}(t_0) + (t - t_0)(v + \varepsilon\mathbb{B}) \subset \Omega(t),$$

and it is enough to use (2.28) and the first inequality in (2.24).

On the other hand, if $t \in]t_0 + k\varrho, t_1]$, then for any projection $\pi(t)$ of $\hat{x}(t - k\varrho)$ onto the set $\Omega(t)$, we have $|\hat{x}(t - k\varrho) - \pi(t)| = d_{\Omega(t)}(\hat{x}(t - k\varrho)) \leq \varrho$, and, from (2.25), it follows that

$$y(t) \in \pi(t) + k\varrho v + \varrho\mathbb{B}. \tag{2.30}$$

Now, since $|\pi(t) - \hat{x}(t_0)| \leq |\hat{x}(t - k\varrho) - \pi(t)| + |\hat{x}(t - k\varrho) - \hat{x}(t_0)| \leq \bar{\varrho} + M\Delta$, from Proposition 2.4.2 and the 2nd inequality in (2.23)

$$\pi(t) + k\varrho v + k\varrho\varepsilon\mathbb{B} = \pi(t) + k\varrho(v + \varepsilon\mathbb{B}) \subset \Omega(t). \tag{2.31}$$

Finally, (2.30) and (2.31) imply that $y(t) + (k\varepsilon - 1)\varrho\mathbb{B} \subset \Omega(t)$. So, the claim follows from (2.24):(ii) and (2.29).

CASE 2: $\varrho > \bar{\varrho}$ and $\delta \leq \Delta$.

Applying Theorem 2.2.4 with $E(.) = \Omega(.)$, there exists a feasible F-trajectory $\bar{x}(.)$ on $[t_0, t_1]$ starting from $\hat{x}(t_0)$. Note that $d_{\Omega(t)}(\bar{x}(t)) = 0$ for all $t \in [t_0, t_1]$. By the Case 1, replacing $\hat{x}(.)$ with $\bar{x}(.)$, it follows that there exists a feasible F-trajectory $x(.)$ on $[t_0, t_1]$ such that $x(t_0) = \hat{x}(t_0)$ and $x(t) \in \text{int}\,\Omega(t)$ for all $t \in]t_0, t_1]$. Hence, by Assumption 2.3.1:(II), we have $\|\hat{x} - x\|_{\infty,[t_0,t_1]} \leq 2M\Delta \leq \beta_2\varrho$, with $\beta_2 = \frac{2M\Delta}{\bar{\varrho}}$.

CASE 3: $\delta > \Delta$.

The above proof implies that in Cases 1 and 2, β_1, β_2 can be taken the same if δ is replaced by any $0 < \delta_1 < \delta$. Define $\tilde{\beta} = \beta_1 \vee \beta_2$ and let $\{[\tau_-^i, \tau_+^i]\}_{i=1}^m$ be a partition of $[t_0, t_1]$ by the intervals with the length at most δ/m. Put $x_0(.) := \hat{x}(.)$. From Cases 1 and 2, replacing $[t_0, t_1]$ by $[\tau_-^1, \tau_+^1]$ and setting $\varrho_0 = \varrho$ we conclude that there exists an F-trajectory $x_1(.)$ on $[\tau_-^1, \tau_+^1] = [t_0, \tau_+^1]$ such that $x_1(t_0) = \hat{x}(t_0)$, $x_1(t) \in \text{int}\,\Omega(t)$ for all $t \in]t_0, \tau_+^1]$, and

$$\|x_1 - x_0\|_{\infty,[\tau_-^1,\tau_+^1]} \leq \tilde{\beta}\varrho_0.$$

Using the Generalized Filippov Existence Theorem, we can extend the trajectory $x_1(.)$ on whole interval $[t_0, t_1]$ so that

$$\|x_1 - x_0\|_{\infty,[t_0,t_1]} \leq e^{\int_{t_0}^{t_1} \varphi(\tau)\,d\tau}\tilde{\beta}\varrho_0 \leq K\tilde{\beta}\varrho_0,$$

where $K := e^{\theta_\varphi(\delta)}$. Repeating recursively the above argument on each time interval $[\tau_-^i, \tau_+^i]$, we conclude that there exists a sequence of F-trajectories $\{x_i(.)\}_{i=1}^m$ on $[t_0, t_1]$, such that:

- $x_i(t_0) = \hat{x}(t_0)$ for all $i = 1, \ldots, m$;
- $x_i(t) \in \text{int}\,\Omega(t)$ for all $t \in]t_0, \tau_+^i]$ and all $i = 1, \ldots, m$;
- $x_j(.)|_{[t_0,\tau_+^{j-1}]} = x_{j-1}(.)$ for all $j = 2, \ldots, m$;

and

$$\|x_i - x_{i-1}\|_{\infty,[t_0,t_1]} \leq K\tilde{\beta}\varrho_{i-1} \qquad \forall i = 1, \ldots, m, \tag{2.32}$$

where

$$\varrho_{i-1} = \max\{\varrho, \sup_{t\in[t_0,t_1]} d_{\Omega(t)}(x_{i-1}(t))\}.$$

Notice that

$$\varrho_i \le \varrho_{i-1} + \|x_i - x_{i-1}\|_{\infty,[t_0,t_1]} \qquad \forall i = 1, \dots, m. \tag{2.33}$$

Taking note of (2.32) and (2.33) we get for all $i = 1, \dots, m$

$$\begin{aligned}
\|x_i - x_{i-1}\|_{\infty,[t_0,t_1]} &\le K\tilde{\beta}(\varrho_{i-2} + \|x_{i-1} - x_{i-2}\|_{\infty,[t_0,t_1]}) \\
&\le K\tilde{\beta}(1 + K\tilde{\beta})\varrho_{i-2} \\
&\le \dots \\
&\le K\tilde{\beta}(1 + K\tilde{\beta})^{i-1}\varrho_0.
\end{aligned}$$

Then, letting $x(.) := x_m(.)$ and observing that $\varrho_0 \le \varrho$, we obtain

$$\begin{aligned}
\|x - \hat{x}\|_{\infty,[t_0,t_1]} &\le \sum_{i=1}^{m} \|x_i - x_{i-1}\|_{\infty,[t_0,t_1]} \\
&\le K\tilde{\beta}\varrho_0 \sum_{i=1}^{m} (1 + K\tilde{\beta})^{i-1} \le \beta_3\varrho,
\end{aligned}$$

where $\beta_3 = (1 + K\tilde{\beta})^m - 1$.

Then all conclusions of the theorem follow with $\beta = \tilde{\beta} \vee \beta_3$. Observe that β depends only on ε, $\hat{\eta}$, M, δ, and on functions $\gamma(.)$ and $\varphi(.)$.

Now, assume $F \neq \text{co } F$. From the first part of the proof, we have that there exist $\beta > 0$ (that does not depend on the reference trajectory $\hat{x}(.)$ on $[t_0, t_1]$) and a co F trajectory $\bar{x}(.) : [t_0, t_1] \to \mathbb{R}^n$, strictly feasible on $]t_0, t_1]$, such that

$$\left\|\bar{x} - \hat{x}\right\|_{\infty,[t_0,t_1]} \le \beta\varrho.$$

Let $(s_i)_i \subset]t_0, t_1]$ with $s_1 = t_1$ be a decreasing sequence such that $s_i \to t_0$. Since $\bar{x}(.)$ is strictly feasible on $]t_0, t_1]$ we can find a sequence of decreasing numbers $(\varepsilon_i)_i \subset]0, \varrho[$ such that $\varepsilon_i \to 0$ and

$$\bar{x}(\sigma) + \varepsilon_i\mathbb{B} \subset \Omega(\sigma) \quad \forall \sigma \in [s_i, t_1], \ \forall i \ge 2. \tag{2.34}$$

Without loss of generality, we can assume that $\varepsilon_i \le \frac{1}{4} \wedge \varrho$ for all $i \in \mathbb{N}^+$. Put $C := e^{\int_{t_0}^{t_1} \varphi(\sigma)d\sigma}$ and define $a_k := \frac{\varepsilon_k}{C^k}$ for all $k \in \mathbb{N}$. Notice that

$$\sum_{k=i}^{+\infty} C^k a_k < \frac{\varepsilon_i}{2} \quad \forall i \ge 2.$$

From Lemma 1.3.3, there exist a sequence of F-trajectories $x_i : [s_i, t_1] \to \mathbb{R}^n$ such that, for all $i \geq 2$, we have $x_i(s_i) = \bar{x}(s_i)$ and

$$\|x_i - \bar{x}\|_{\infty,[s_i,t_1]} \leq a_i. \tag{2.35}$$

For each integer $j \geq 2$, we construct an F-trajectory $y_j : [s_j, t_1] \to \mathbb{R}^n$ as follows. We put $y_2(.) := x_2(.)$ and for all $j > 2$:

- $y_j(.)$ is the restriction of $x_j(.)$ on $]s_j, s_{j-1}]$;
- for all $1 \leq k \leq j-2$, $y_j(.)$ restricted to $]s_{j-k}, s_{j-k-1}]$ is an F-trajectory with initial state $y_j(s_{j-k})$, obtained by applying the Generalized Filippov Existence Theorem with reference trajectory $y_{j-1}(.)$.

Now fix an integer $j > 2$. From the Generalized Filippov Existence Theorem and since $x_i(s_i) = \bar{x}(s_i)$, for any $2 \leq i < j$

$$\|y_j - x_i\|_{\infty,[s_i,s_{i-1}]} \leq C\left|y_j(s_i) - \bar{x}(s_i)\right|,$$
$$\|y_j' - x_i'\|_{1,[s_i,s_{i-1}]} \leq C\left|y_j(s_i) - \bar{x}(s_i)\right|.$$

From these relations and (2.35) it follows that for each $2 \leq i < j$ and any $l \in \mathbb{N}^+$

$$\|y_j - \bar{x}\|_{\infty,[s_i,s_{i-1}]} \leq \sum_{k=i}^{j} C^{k-i} a_k, \tag{2.36}$$

$$\|y_{j+l}' - y_j'\|_{1,[s_i,s_{i-1}]} \leq 2 \sum_{k=i+1}^{j+l} C^{k-i} a_k. \tag{2.37}$$

Notice that $y_j(s_j) = \bar{x}(s_j)$ for any $j \geq 2$. Hence we can extend each F-trajectory y_j as an co F trajectory to whole interval $[t_0, t_1]$, by setting $y_j(\sigma) = \bar{x}(\sigma)$ for $\sigma \in [t_0, s_j]$. Since the sequence of trajectories $(y_i)_i$ have initial value $\hat{x}(t_0)$ and owing the sub-linear growth of F, taking a subsequence and keeping the same notation, we have

$$\exists \text{ co } F\text{-trajectory } x(.) : \ y_i \to x \text{ in } C([t_0, t_1]; \mathbb{R}^n), \text{ with } x(t_0) = \hat{x}(t_0).$$

We conclude to show that $x(.)$ satisfy all the conclusions with β replaced by $\beta + 1$. Indeed, due to (2.37), for each $k \geq 2$ the F-trajectories trajectories $(y_i)_i$, restricted to $\left[s_k, s_{k-1}\right]$, forms a Cauchy sequence on $W^{1,1}([s_k, s_{k-1}]; \mathbb{R}^n)$. So, it follows that the limiting co F-trajectory $x(.)$ is an F-trajectory and, since $\varepsilon_i \leq \varrho$ for all $i \geq 2$,

$$\|\hat{x} - x\|_{\infty,[t_0,t_1]} \leq \left\|\hat{x} - \bar{x}\right\|_{\infty,[t_0,t_1]} + \|x - \bar{x}\|_{\infty,[t_0,t_1]} \leq (\beta + 1)\varrho.$$

Moreover, notice that $x(.)$ is strictly feasible on $]t_0, t_1]$. Indeed, consider $\sigma \in]t_0, t_1]$. We have $\sigma \in]s_i, s_{i-1}]$ for some $i \geq 2$. From (2.36), (2.35), and (2.34) we get

$$y_j(\sigma) \in \bar{x}(\sigma) + \frac{\varepsilon_i}{2}\mathbb{B} \subset \operatorname{int}\Omega(\sigma) \quad \text{for all } j \geq i.$$

Since the $(y_j)_j$ converge uniformly to x,

$$x(\sigma) \in \bar{x}(\sigma) + \frac{\varepsilon_i}{2}\mathbb{B} \subset \operatorname{int}\Omega(\sigma).$$

This concludes our proof. □

2.3.1 Neighboring Estimates for Autonomous Tubes

In the autonomous case, the time-invariant structure of the dynamics allows for more refined analysis of neighboring trajectory behavior. This section specializes in general neighboring feasible trajectory estimates to autonomous systems, where we can exploit the invariance properties under mild assumptions on controllability, rather than those used in the previous section. We deal with uniform linear estimates on intervals $I = [t_0, t_1], 0 \leq t_0 < t_1$, for a general state constrained differential inclusion of the form

$$\begin{cases} x'(t) \in F(t, x(t)) & \text{a.e. } t \in I, \\ x(t) \in \Omega & \forall t \in I, \end{cases}$$

where $F: \mathbb{R}_0^+ \times \mathbb{R}^n \rightsquigarrow \mathbb{R}^n$ is a given set-valued map and $\Omega \subset \mathbb{R}^n$ is a closed nonempty set. We consider the following assumptions on $F(.,.)$:

Assumption 2.3.4 We consider the following condition on $F(.,.)$:

(I) F has closed, nonempty values, a sub-linear growth, and $F(., x)$ is Lebesgue measurable for all $x \in \mathbb{R}^n$;
(II) there exists $\varphi \in \mathcal{L}_{\text{loc}}$ such that $F(t, .)$ is $\varphi(t)$-Lipschitz continuous for a.e. $t \in [0, +\infty[$;
(III) there exists $q \in \mathcal{L}_{\text{loc}}$ suchthat $F(t, x) \subset q(t)\mathbb{B}$ for all $x \in \operatorname{bdr}\Omega$, and a.e. $t \in \mathbb{R}_0^+$.

Assumption 2.3.5 (*Inward Pointing Condition*) There exist $\eta > 0$, $r > 0$, and $M \geq 0$ such that: for a.e. $t \in \mathbb{R}_0^+$, all $y \in \operatorname{bdr}\Omega + \eta\mathbb{B}$, and all

$$v \in F(t, y) \cap \{p \in \mathbb{R}^n \mid \exists n \in N_\Omega(y; \eta),\ \langle p, n \rangle \geq 0\},$$

there exists $w \in F(t, y) \cap B(v, M)$ such that

$$w,\ w - p \in \{p \in \mathbb{R}^n \mid \langle p, n \rangle \leq -r,\ \forall n \in N_\Omega(y; \eta)\},$$

where

$$N_\Omega(x;\eta) := \{n \in \mathbb{S} \mid n \in \text{cl co}\, N_\Omega(y),\ y \in (\text{bdr}\,\Omega) \cap B(x,\eta)\}. \tag{2.38}$$

Theorem 2.3.6 (Neighboring Feasible Estimates for Autonomous Tubes) *Consider Assumptions 2.3.4 and 2.3.5. Then all the conclusions of Theorem 2.3.3 hold true for the constant tube* $\Omega(.) \equiv \Omega$.

Proof Fix $\delta > 0$ and let $k > 0$, $\Delta > 0$ be such that $k > 1/r$, and

$$(i)\ 8\Delta M \le \eta,\ \ (ii)\ 2e^{\theta_\varphi(\Delta)}\theta_\varphi(\Delta)M < r,\ \ (iii)\ 2e^{\theta_\varphi(\Delta)}\theta_\varphi(\Delta)Mk < rk - 1. \tag{2.39}$$

We notice that, from Assumption 2.3.4:(II), (III), for any $\alpha > 0$ there exists $q_\alpha \in \mathcal{L}_{\text{loc}}$ such that $F(t,x) \subset q_\alpha(t)\mathbb{B}$ for a.e. $t > 0$ and all $x \in \text{bdr}\,\Omega + \alpha\mathbb{B}$. So, by the proof of Theorem 2.3.3, without loss of generality we can replace M with a suitable greater constant and suppose that $\delta \le \Delta$, $\int_J q_\eta \le M$ for any $J \subset \mathbb{R}_0^+$ with $\mu_{\mathcal{L}}(J) = \delta$. Moreover, by (2.39):$(i)$ we have that if $\hat{x}(t_0) \in \Omega \backslash (\text{bdr}\,\Omega + \frac{\eta}{4}\mathbb{B})$ then the conclusion follows taking $x(.) = \hat{x}(.)$. Then we suppose that $\hat{x}(t_0) \in (\text{bdr}\,\Omega + \frac{\eta}{4}\mathbb{B}) \cap \Omega$. Let us denote the nonsmooth orientated distance $Dist_{\text{bdr}\,\Omega}(.)$ from the boundary of Ω by

$$Dist_{\text{bdr}\,\Omega}(.) := d_\Omega(.) - d_{\Omega^c}(.)$$

and define the measurable set

$$\Omega_+ := \left\{ s \in [t_0,t_1] \cap J \,\middle|\, \exists x \in (\text{bdr}\,\Omega) + B(\hat{x}(s),\eta), \exists n \in N_\Omega^C(x) \cap \mathbb{S},\ \langle n, \hat{x}'(s)\rangle \ge 0 \right\},$$

where $J := \{s \in [t_0,t_1] \mid \hat{x}'(s) \text{ exists}\}$. We show the stated uniform estimate by distinguish the cases $\mu_{\mathcal{L}}(\Omega_+) = 0$ or $\mu_{\mathcal{L}}(\Omega_+) > 0$.

CASE 1: $\mu_{\mathcal{L}}(\Omega_+) = 0$.

Let $t \in]t_0,t_1]$. Then applying the Main Value Theorem, for some $z(t) \in [\hat{x}(t_0), \hat{x}(t)]$ and $\xi(t) \in \partial^C Dist_{\text{bdr}\,\Omega}(z(t))$, we have

$$Dist_{\text{bdr}\,\Omega}(\hat{x}(t)) = Dist_{\text{bdr}\,\Omega}(\hat{x}(t_0)) + \langle \xi(t), \hat{x}(t) - \hat{x}(t_0)\rangle.$$

From Lemma 1.2.1, any vector of the form $v = (x - \tilde{x})/d_E(x)$, with $\tilde{x}$ in the projection set of $x \ne \tilde{x}$ onto E, is a proximal normal to E at $\tilde{x}$, and in particular it lays in cl co $N_E(x)$. Hence, from (1.3) and Lemma 1.2.1, we have that there exist $\lambda_i > 0$, y_i in the set of projections of $z(t)$ onto bdr Ω, and $\xi_i \in N_\Omega^C(y_i) \cap \mathbb{S}$, for $i = 1, \ldots, N$, with $1 \le N \le n+1$, such that $\sum_{i=1}^N \lambda_i = 1$ and $\xi(t) = \sum_{i=1}^N \lambda_i \xi_i$. Moreover, for any $s \in [t_0,t_1]$

$$\begin{aligned}
|y_i - \hat{x}(s)| &\le |y_i - z(t)| + |z(t) - \hat{x}(s)| \\
&\le d_{\text{bdr}\,\Omega}(z(t)) + |\hat{x}(t_0) - \hat{x}(s)| \\
&\le d_{\text{bdr}\,\Omega}(\hat{x}(t_0)) + |z(t) - \hat{x}(t_0)| + |\hat{x}(t_0) - \hat{x}(s)| \\
&\le \frac{\eta}{4} + 2\frac{\eta}{8} = \eta.
\end{aligned}$$

Hence, $Dist_{\text{bdr}\,\Omega}(\hat{x}(t)) = Dist_{\text{bdr}\,\Omega}(\hat{x}(t_0)) + \sum_{i=1}^{N} \lambda_i \int_{t_0}^{t} \langle \xi_i, \hat{x}'(s)\rangle ds < 0$ for all $t \in]t_0, t_1]$, and $x(.) = \hat{x}(.)$ satisfies the conclusions.

CASE 2: $\mu_{\mathscr{L}}(\Omega_+) > 0$.

Applying the Measurable Selection Theorem, let $w : \Omega_+ \to \mathbb{R}^n$ be the Lebesgue measurable function satisfying $w(t) \in F(t, \hat{x}(t))$ for a.e. $t \in \Omega_+$, as in Assumption 2.3.5. Define

$$\tau := \begin{cases} t_1 & \text{if } \mu_{\mathscr{L}}(\Omega_+) \le k\rho, \\ \inf\{t \in]t_0, t_1] \mid \mu_{\mathscr{L}}(\Omega_+ \cap [t_0, t]) = k\rho\} & \text{otherwise,} \end{cases}$$

and keep $y : [t_0, t_1] \to \mathbb{R}^n$ the arc satisfying $y(t_0) = \hat{x}(t_0)$ and

$$y'(t) := \begin{cases} w(t) & \text{if } t \in \Omega_+ \cap [t_0, \tau], \\ \hat{x}'(t) & \text{if } t \in [t_0, t_1] \setminus (\Omega_+ \cap [t_0, \tau]) \cap J. \end{cases} \tag{2.40}$$

So, we have for all $t \in [t_0, t_1]$

$$\begin{aligned} &\left\| \hat{x} - y \right\|_{W^{1,1},[t_0,t]} \\ &= \left| y(t_0) - \hat{x}(t_0) \right| + \int_{t_0}^{t} \left| y'(s) - \hat{x}'(s) \right| ds \\ &= \int_{t_0}^{\tau \wedge t} \left| w(s) - \hat{x}'(s) \right| \chi_{\Omega_+ \cap [t_0,\tau]}(s) ds \le 2M \mu_{\mathscr{L}}(\Omega_+ \cap [t_0, \tau \wedge t]). \end{aligned} \tag{2.41}$$

Applying the Generalized Filippov Existence Theorem, we have that there exists an F-trajectory $x(.)$ on $[t_0, t_1]$ such that $x(t_0) = y(t_0)$ and

$$\| y - x \|_{W^{1,1},[t_0,t]} \le e^{\int_{t_0}^{t} \varphi(\tau)\, d\tau} \int_{t_0}^{t} d_{F(s,y(s))}(y'(s)) ds \tag{2.42}$$

for all $t \in [t_0, t_1]$. Now, since for any $s \in [t_0, t_1]$, $y(s) = \hat{x}(t_0) + \int_{t_0}^{s} y'(\xi)\, d\xi = \hat{x}(s) + \int_{t_0}^{s} \left(y'(\xi) - \hat{x}'(\xi) \right) d\xi$, then for a.e. $s \in [t_0, t_1]$ we have $F(s, \hat{x}(s)) \subset F(s, y(s)) + \varphi(s) \int_{t_0}^{s} \left| y'(\xi) - \hat{x}'(\xi) \right| d\xi\, \mathbb{B}$. So, taking note of (2.40), we have for a.e. $s \in [t_0, t_1]$

$$\begin{aligned} d_{F(s,y(s))}(y'(s)) &\le d_{F(s,\hat{x}(s))}(y'(s)) + \varphi(s) \int_{t_0}^{s} \left| y'(\xi) - \hat{x}'(\xi) \right| d\xi \\ &= \varphi(s) \int_{t_0}^{s} \left| y'(\xi) - \hat{x}'(\xi) \right| d\xi. \end{aligned} \tag{2.43}$$

Using (2.43) and (2.41), it follows that $d_{F(s,y(s))}(y'(s)) \le 2\varphi(s) M \mu_{\mathscr{L}}(\Omega_+ \cap [t_0, \tau \wedge s])$ for a.e. $s \in [t_0, t_1]$. Hence, we obtain the estimates for all $t \in [t_0, t_1]$

$$\int_{t_0}^{t} d_{F(s,y(s))}(y'(s))ds \leq 2\theta_\varphi(\Delta)M\mu_{\mathscr{L}}(\Omega_+ \cap [t_0, \tau \wedge t]).$$

Thus, by (2.42), we have $\|y - x\|_{W^{1,1},[t_0,t]} \leq 2e^{\theta_\varphi(\Delta)}\theta_\varphi(\Delta)M\mu_{\mathscr{L}}(\Omega_+ \cap [t_0, \tau \wedge t])$ for all $t \in [t_0, t_1]$, and, using (2.41), we get

$$\left\|\hat{x} - x\right\|_{W^{1,1},[t_0,t_1]} \leq \beta\rho \ \text{ with } \beta = 2M\left(e^{\theta_\varphi(\Delta)}\theta_\varphi(\Delta) + 1\right)k.$$

To conclude the proof, we show that

$$x(t) \in \operatorname{int}\Omega \quad \forall t \in]t_0, t_1].$$

Consider $t \in]t_0, \tau]$. Then, applying the Main Value Theorem, we have

$$Dist_{\text{bdr}\,\Omega}(y(t)) = Dist_{\text{bdr}\,\Omega}(y(t_0)) + \langle \xi(t), y(t) - y(t_0)\rangle$$

for some $z(t) \in [y(t_0), y(t)]$ and $\xi(t) \in \partial^C Dist_{\text{bdr}\,\Omega}(z(t))$. Now, we have that there exist $\lambda_i > 0$, y_i in the set of projections of $z(t)$ onto bdr Ω, and $\xi_i \in N^C_\Omega(y_i) \cap \mathbb{S}$, for $i = 1, \ldots, N$, with $1 \leq N \leq n+1$, such that $\sum_{i=1}^N \lambda_i = 1$ and $\xi(t) = \sum_{i=1}^N \lambda_i \xi_i$. Since for all $s \in [t_0, t_1]$ and all $i = 1, \ldots, N$, $\left|y_i - \hat{x}(s)\right| \leq |y_i - z(t)| + \left|z(t) - \hat{x}(t_0)\right| + \left|\hat{x}(s) - \hat{x}(t_0)\right|$ and

$$\begin{aligned} |y_i - z(t)| &\leq d_{\text{bdr}\,\Omega}(y(t)) + |d_{\text{bdr}\,\Omega}(z(t)) - d_{\text{bdr}\,\Omega}(y(t))| \\ &\leq d_{\text{bdr}\,\Omega}(y(t)) + |y(t) - y(t_0)| \\ &\leq d_{\text{bdr}\,\Omega}(y(t_0)) + 2\,|y(t) - y(t_0)|, \end{aligned}$$

we have that $\left|y_i - \hat{x}(s)\right| \leq \frac{\eta}{4} + 2\frac{\eta}{8} + 2\frac{\eta}{8} < \eta$. It follows that

$$\begin{aligned} \langle \xi(t), y(t) - y(t_0)\rangle &= \int_{[t_0,t]\setminus\Omega_+} \langle \xi(t), \hat{x}'(s)\rangle ds + \int_{[t_0,t]\cap\Omega_+} \langle \xi(t), w(s)\rangle ds \\ &= \sum_{i=1}^N \int_{[t_0,t]\setminus\Omega_+} \langle \xi_i, \hat{x}'(s)\rangle ds + \sum_{i=1}^N \int_{[t_0,t]\cap\Omega_+} \lambda_i \langle \xi_i, w(s)\rangle ds \\ &\leq -r\mu_{\mathscr{L}}(\Omega_+ \cap [t_0, t]). \end{aligned} \tag{2.44}$$

Hence, using (2.39):(ii),

$$\begin{aligned} Dist_{\text{bdr}\,\Omega}(x(t)) &\leq |x(t) - y(t)| + Dist_{\text{bdr}\,\Omega}(y(t)) \\ &\leq \left(2e^{\theta_\varphi(\Delta)}\theta_\varphi(\Delta)M - r\right)\mu_{\mathscr{L}}(\Omega_+ \cap [t_0, t]) \leq 0. \end{aligned} \tag{2.45}$$

The equality in (2.44) occurs only if $\mu_{\mathscr{L}}([t_0, t]\setminus\Omega_+) = 0$, and in that case it follows that $\mu_{\mathscr{L}}(\Omega_+ \cap [t_0, t])$ has positive measure. It follows that the inequality in (2.45) is strict.

Consider $t \in]\tau, t_1]$. By the Main Value Theorem, for some $z(t) \in [\hat{x}(t), y(t)]$ and $\xi(t) \in \partial^C Dist_{\text{bdr}\,\Omega}(z(t))$, we have $Dist_{\text{bdr}\,\Omega}(y(t)) = Dist_{\text{bdr}\,\Omega}(\hat{x}(t)) + \langle \xi(t), y(t) - \hat{x}(t)\rangle$. Furthermore, arguing as in the previous case, consider $\lambda_i > 0$, y_i in the set of projections of $z(t)$ onto bdr Ω, and $\xi_i \in N^C_\Omega(y_i) \cap \mathbb{S}$, for $i = 1, \dots, N$, with $1 \le N \le n+1$, such that $\sum_{i=1}^N \lambda_i = 1$ and $\xi(t) = \sum_{i=1}^N \lambda_i \xi_i$. We notice that for all $i = 1, \dots, N$ and $s \in]\tau, t_1]$, $\left|y_i - \hat{x}(s)\right| \le |y_i - z(t)| + \left|z(t) - \hat{x}(s)\right|$ and

$$\begin{aligned}
&\left|z(t) - \hat{x}(s)\right| \\
&\le \frac{1}{2}(|z(t) - y(t)| + |y(t) - y(t_0)| + \left|\hat{x}(t_0) - \hat{x}(s)\right| + \left|\hat{x}(s) - \hat{x}(t)\right| \\
&\qquad + \left|\hat{x}(t) - z(t)\right|) \\
&\le \left|\hat{x}(t) - y(t)\right| + \frac{1}{2}\left(|y(t) - y(t_0)| + \left|\hat{x}(t_0) - \hat{x}(s)\right| + \left|\hat{x}(s) - \hat{x}(t)\right|\right).
\end{aligned}$$

Now, since $z(t) = ay(t) + (1-a)\hat{x}(t)$, where $a \in [0, 1]$, we have that $|y_i - z(t)| \le Dist_{\text{bdr}\,\Omega}(\hat{x}(t_0)) + \left|\hat{x}(t_0) - z(t)\right| \le \frac{\eta}{4} + |y(t) - y(t_0)| + \left|\hat{x}(t) - \hat{x}(t_0)\right|$. Summing up we obtain $\left|y_i - \hat{x}(s)\right| \le \left(\frac{\eta}{4} + 2\frac{\eta}{8}\right) + 2\frac{\eta}{8} + \frac{1}{2}3\frac{\eta}{8} < \eta$. Then

$$\begin{aligned}
\langle \xi(t), y(t) - \hat{x}(t)\rangle &= \int_{t_0}^{t} \langle \xi(t), y'(s) - x'(s)\rangle ds \\
&= \int_{\Omega_+ \cap [t_0, \tau]} \langle \xi(t), w(s) - x'(s)\rangle ds \\
&= \sum_{i=1}^{N} \int_{\Omega_+ \cap [t_0, \tau]} \lambda_i \langle \xi_i, w(s) - x'(s)\rangle ds \\
&\le -r\mu_{\mathscr{L}}(\Omega_+ \cap [t_0, \tau]).
\end{aligned}$$

Finally, by (2.39):(iii),

$$\begin{aligned}
Dist_{\text{bdr}\,\Omega}(x(t)) &\le |x(t) - y(t)| + Dist_{\text{bdr}\,\Omega}(y(t)) \\
&\le 2e^{\theta_\varphi(\Delta)}\theta_\varphi(\Delta)Mk\rho + Dist_{\text{bdr}\,\Omega}(y(t)) \\
&\le Dist_{\text{bdr}\,\Omega}(\hat{x}(t)) + \left(2e^{\theta_\varphi(\Delta)}\theta_\varphi(\Delta)M - r\right)k\rho \\
&\le (1 + (2e^{\theta_\varphi(\Delta)}\theta_\varphi(\Delta)M - r)k)\rho < 0,
\end{aligned}$$

end the proof is complete. □

2.4 Sufficient Conditions for Controllability

Having established viability results, it is natural to ask when the system is actually controllable–that is, when there exist controls that allow reaching specific objectives while respecting the constraints. In this section, we present sufficient geometric conditions for controllability, based on the inward pointing condition applied to the infinite horizon case with time-dependent state constraints.

We consider the following assumptions.

Assumption 2.4.1 Let $\theta \in]0, 1]$ and

$$\Omega(t) := \bigcap_{i=1}^{m} \Omega_i(t), \quad \Omega_i(t) := \{x \in \mathbb{R}^n \mid h_i(t, x) \leq 0\},$$

where $h_i : \mathbb{R}_0^+ \times \mathbb{R}^n \to \mathbb{R}$ be m real-valued functions satisfying for any $i = 1, \ldots, m$:

- $h_i(., x)$ is measurable for any x;
- $h_i(t, .)$ is $\Gamma^{1,\theta}$ regular, uniformly wrt t.

The result below states a geometric result for a inward pointing condition on infinite horizon wrt the constraints $\Omega(t)$ and the dynamics.

Proposition 2.4.2 *Consider Assumption 2.4.1. Let $F : \mathbb{R} \times \mathbb{R}^n \rightsquigarrow \mathbb{R}^n$ be a set-valued map with nonempty closed values satisfying:*

$$\begin{cases} \exists M \geq 0, \exists \varphi > 0 : \\ (a) \ \sup\{|v| \mid v \in F(t, x), \ t \in \mathbb{R}_0^+, x \in \operatorname{bdr} \Omega(t)\} \leq M; \\ (b) \ F(t, .) \text{ is } \varphi\text{-Lipschitz continuous for any } t \geq 0. \end{cases}$$

Assume that

$$\begin{cases} \exists \delta > 0, \ \exists r > 0\text{: } \forall t \in \mathbb{R}_0^+, \ \forall x \in \operatorname{bdr} \Omega(t), \\ \exists v \in \operatorname{co} F(t, x) \text{ satisfying} \\ \langle \nabla h_i(t, x), v \rangle \leq -r \qquad \forall i \in \bigcup_{z \in B(x,\delta)} I(z), \end{cases} \tag{2.46}$$

where $I(z) = \{i \in I \mid z \in \operatorname{bdr} \Omega_i(t)\}$ and $I := \{1, \ldots, m\}$. Then Assumption 2.3.2 hold true.

Proof Let us set $J(x) := \bigcup_{z \in B(x,\delta)} I(z)$ for all $x \in \operatorname{bdr} \Omega(s)$ and $s \geq 0$. Fix $t \in \mathbb{R}_0^+$, $x \in \operatorname{bdr} \Omega(t)$, and $v \in \operatorname{co} F(t, x)$ satisfying $\langle \nabla h_i(t, x), v \rangle \leq -r$ for all $i \in J(x)$. Pick

$$k > \max_{i \in I} \sup_{x \neq y} \frac{|\nabla h_i(t, x) - \nabla h_i(t, y)|}{|x - y|^\theta},$$

$$L > \max_{i \in I} \sup_{x \in \mathbb{R}^n} |\nabla h_i(t, x)|.$$

We proceed by steps.

STEP 1: We claim that there exists $\tilde{\eta} > 0$, not depending on (t, x), such that for all $y \in B(x, \tilde{\eta})$ we can find $w \in \text{co}\, F(t, y)$, with $|w - v| \leq r/4L$, satisfying for all $i \in J(x)$,

$$\langle \nabla h_i(t, y), w \rangle \leq -r/2.$$

Indeed, for all $i \in J(x)$ and $y \in B(x, \sqrt[\theta]{r/4kM})$ we have

$$\begin{aligned}\langle \nabla h_i(t, y), v \rangle &= \langle \nabla h_i(t, y) - \nabla h_i(t, x), v \rangle + \langle \nabla h_i(t, x), v \rangle \\ &\leq kM|y - x| - r \leq -\frac{3r}{4}\end{aligned}$$

and for all $w \in \mathbb{R}^n$ such that $|w - v| \leq r/4L$

$$\begin{aligned}\langle \nabla h_i(t, y), w \rangle &= \langle \nabla h_i(t, y), w - v \rangle + \langle \nabla h_i(t, y), v \rangle \\ &\leq L|w - v| - 3r/4 \leq -\frac{r}{2}.\end{aligned}$$

Since $F(t, .)$ is φ-Lipschitz continuous, there exists $w \in \text{co}\, F(t, y)$ such that $|w - v| \leq r/4L$ whenever $|y - x| \leq r/4\varphi L$. So the claim follows with $\tilde{\eta} = \min\{r/4\varphi L, \sqrt[\theta]{r/4kM}\}$.

STEP 2: We claim that there exists $\tilde{\varepsilon} > 0$, not depending on (t, x), such that for all $y \in B(x, \tilde{\eta})$ we can find $w \in \text{co}\, F(t, y)$ such that

$$\langle \nabla h_i(t, z), \tilde{w} \rangle \leq -r/4 \qquad \forall z \in B(y, \tilde{\varepsilon}),\ \forall \tilde{w} \in B(w, \tilde{\varepsilon}),\ \forall i \in J(x).$$

Indeed, let $y \in B(x, \tilde{\eta})$ and $w \in \text{co}\, F(t, y)$ be as in Step 1. Then for any $\tilde{w} \in \mathbb{R}^n$ such that $|\tilde{w} - w| \leq r/8L$ and for all $i \in J(x)$ and $z \in \mathbb{R}^n$,

$$\begin{aligned}&\langle \nabla h_i(t, z), \tilde{w} \rangle \\ &= \langle \nabla h_i(t, z) - \nabla h_i(t, y), \tilde{w} \rangle + \langle \nabla h_i(t, y), \tilde{w} - w \rangle + \langle \nabla h_i(t, y), w \rangle \\ &\leq k(M + r/4L + r/8L)|z - y| + r/8 - r/2.\end{aligned}$$

So the claim follows with $\tilde{\varepsilon} = \min\{k^{-1}(M + r/2L)^{-1}r/8,\ r/8L\}$.

STEP 3: We prove that there exist $\eta > 0$, $\varepsilon > 0$, not depending on (t, x), such that for all $y \in B(x, \eta) \cap \Omega(t)$ we can find $w \in \text{co}\, F(t, y)$ satisfying

$$z + \tau \tilde{w} \in \Omega(t) \qquad \forall z \in B(y, \varepsilon) \cap \Omega(t),\ \forall \tilde{w} \in B(w, \varepsilon),\ \forall 0 \leq \tau \leq \varepsilon.$$

Let $y \in B(x, \tilde{\eta}) \cap \Omega(t)$ and $w \in \text{co}\, F(t, y)$ be as in Step 2. Then, by the Mean Value Theorem for real valued functions, for any $\tau \geq 0$, any $z \in B(y, \tilde{\varepsilon}) \cap \Omega(t)$, any $\tilde{w} \in B(w, \tilde{\varepsilon})$, and any $i \in J(x)$ there exists $\sigma_\tau \in [0, 1]$ such that

$$\begin{aligned} h_i(t, z + \tau\tilde{w}) &= h_i(t, z) + \tau\langle \nabla h_i(t, z + \sigma_\tau \tau\tilde{w}), \tilde{w}\rangle \\ &\leq \tau\langle \nabla h_i(t, z), \tilde{w}\rangle + k(M + r/4L + \tilde{\varepsilon})^2\tau^2 \\ &\leq -\frac{r\tau}{4} + k(M + r/4L + \tilde{\varepsilon})^2\tau^2. \end{aligned}$$

Choosing $\eta \in]0, \tilde{\eta}]$ and $\varepsilon \in]0, \tilde{\varepsilon}]$ such that $\eta + \varepsilon(M + r/4L + \varepsilon) \leq \delta$ and $\varepsilon \leq k^{-1}(M + r/4L + \tilde{\varepsilon})^{-2} r/4$, it follows that for all $z \in B(y, \varepsilon) \cap \Omega(t)$, $\tilde{w} \in B(w, \varepsilon)$, and all $0 \leq \tau \leq \varepsilon$

$$z + \tau\tilde{w} \in B(x, \delta) \tag{2.47}$$

and

$$h_i(t, z + \tau\tilde{w}) \leq 0 \qquad \forall i \in J(x). \tag{2.48}$$

Furthermore, by (2.47) and since $B(x, \delta) \subset \Omega_j(t)$ for all $j \in I \backslash J(x)$, we have for all $z \in B(y, \varepsilon) \cap \Omega(t)$, $\tilde{w} \in B(w, \varepsilon)$, and all $0 \leq \tau \leq \varepsilon$

$$h_i(t, z + \tau\tilde{w}) \leq 0 \qquad \forall i \in I \backslash J(x). \tag{2.49}$$

The conclusion follows from (2.48) and (2.49). □

Remark 2.4.3 (i) From the definition of Boulingad tangent cone, it follows that condition Assumption 2.3.2 implies condition (2.7). So all the conclusions of Theorem 2.2.4 follows under the stronger assumption (2.7).
(ii) We notice that, whenever Proposition 2.4.2 applies, then condition (2.46) on $F(t, x) = \{f(t, x, u) \mid u \in U(t)\}$ ensure the nontriviality intersection (2.7) for $E(t) = \Omega(t)$.

2.5 Lipschitz Continuity

The regularity of the value function represents a crucial aspect in optimal control theory, as it determines the validity of optimality conditions and the numerical stability of algorithms. In this section, we apply the results from previous sections to establish Lipschitz continuity of the value function for a class of infinite horizon optimal control problems subject to state constraints.

Let us consider the problem $(\mathcal{P})$ stated in Sect. 2.1.

Assumption 2.5.1 We take the following assumptions on f and L:

(I) for all $x \in \mathbb{R}^n$ the mappings $f(., x, .)$, $L(., x, .)$ are Lebesgue-Borel measurable;
(II) there exists $\alpha > 0$ such that f and L are bounded functions on

$$\{(t, x, u) \mid t \geq 0,\ x \in (\text{bdr}\ \Omega(t) + \alpha\mathbb{B}),\ u \in U(t)\};$$

(III) for all $(t, x) \in \mathbb{R}_0^+ \times \mathbb{R}^n$ the set

$$\{(f(t, x, u), L(t, x, u)) \mid u \in U(t)\}$$

is closed;

(IV) there exist $c \in L^1_{\text{loc}}(\mathbb{R}_0^+; \mathbb{R}_0^+)$ and $k \in \mathcal{L}_{\text{loc}}$ such that for any $t \in \mathbb{R}_0^+$, $x, y \in \mathbb{R}^n$, and $u \in U(t)$ such that:

(IV.1) $|f(t, x, u) - f(t, y, u)| + |L(t, x, u) - L(t, y, u)| \le k(t)|x - y|$;

(IV.2) $|f(t, x, u)| + |L(t, x, u)| \le c(t)(1 + |x|)$;

(V) there exist $\tilde{\eta} > 0$ and $\gamma \in \mathcal{L}_{\text{loc}}$ such that

$$t \rightsquigarrow \{(f(t, x, u), L(t, x, u)) \mid u \in U(t)\}$$

is γ-left absolutely continuous, uniformly wrt bdr $\Omega + \tilde{\eta}\mathbb{B}$;

(VI) $\limsup_{t \to +\infty} \frac{1}{t} \int_0^t (c(s) + k(s))\, ds < +\infty$.

We consider, for any $\lambda > 0$, the relaxed infinite horizon state constrained problem

$$\text{minimize} \quad \int_t^{+\infty} e^{-\lambda s} L^\star(s, x(s), w(s)) ds \tag{$\mathcal{P}^\star$}$$

over all feasible trajectory-control pairs (x, u) satisfying

$$\begin{cases} x'(s) = f^\star(s, x(s), u(s)) & \text{a.e. } s \in \mathbb{R}_t^+, \\ x(t) = \bar{x}, \\ w(s) \in W(s) & \text{a.e. } s \in \mathbb{R}_t^+, \\ x(s) \in \Omega(s) & \forall s \in \mathbb{R}_t^+, \end{cases}$$

where

$$\begin{aligned} W &: \mathbb{R}_0^+ \rightsquigarrow \mathbb{R}^{(n+1)m} \times \mathbb{R}^{n+1}, \\ f^\star &: \mathbb{R}_0^+ \times \mathbb{R}^n \times \mathbb{R}^{(n+1)m} \times \mathbb{R}^{n+1} \to \mathbb{R}^n, \\ L^\star &: \mathbb{R}_0^+ \times \mathbb{R}^n \times \mathbb{R}^{(n+1)m} \times \mathbb{R}^{n+1} \to \mathbb{R}, \end{aligned}$$

are defined as follows: for all $s \ge 0$

$$W(s) := (\times_{i=0}^n U(s)) \times \{(\alpha_0, \dots, \alpha_n) \in \mathbb{R}^{n+1} \mid \sum_{i=0}^n \alpha_i = 1,\ \alpha_i \ge 0\ \forall i\},$$

and for all $s \ge 0$, $x \in \mathbb{R}^n$, and $w = (u_0, \dots, u_n, \alpha_0, \dots, \alpha_n) \in \mathbb{R}^{(n+1)m} \times \mathbb{R}^{n+1}$

$$f^\star(s, x, w) := \sum_{i=0}^n \alpha_i f(s, x, u_i), \quad L^\star(s, x, w) := \sum_{i=0}^n \alpha_i L(s, x, u_i).$$

Remark 2.5.2 (i) For control systems, the condition Assumption 2.3.2 take the following form: for some $\varepsilon > 0$ and $\eta > 0$ and every $t \in \mathbb{R}_0^+$ and $x \in (\text{bdr}\,\Omega(t) + \eta\mathbb{B}) \cap \Omega(t)$ there exist $\{\alpha_i\}_{i=0}^n \subset [0,1]$, with $\sum_{i=0}^n \alpha_i = 1$, and $\{u_i\}_{i=0}^n \subset U(t)$ satisfying

$$\{y + [0,\varepsilon](\sum_{i=0}^{n} \alpha_i f(t,x,u_i) + \varepsilon\mathbb{B}) \mid y \in (x + \varepsilon\mathbb{B}) \cap \Omega(t)\} \subset \Omega(t).$$

(ii) If there exist $\tilde{\eta} > 0$, γ, $\tilde{\gamma} \in \mathcal{L}_{\text{loc}}$, and $k \geq 0$ such that (f, L) is γ-left absolutely continuous, uniformly wrt $(\text{bdr}\,\Omega + \tilde{\eta}\mathbb{B}) \times \mathbb{R}^m$, $U(.)$ is $\tilde{\gamma}$-left absolutely continuous, and $f(t,x,.)$ is k-Lipschitz continuous for all $t \in \mathbb{R}_0^+$, $x \in (\text{bdr}\,\Omega(t) + \tilde{\eta}\mathbb{B})$, then Assumption 2.5.1:(V) holds true.

Definition 2.5.3 We denote by $V : Q_\Omega \to [-\infty, +\infty]$ the value function of the infinite horizon control problem ($\mathcal{P}$), i.e.

$$V(t,\bar{x}) = \inf\{\int_t^{+\infty} e^{-\lambda s} L(s, x(s), u(s))ds \mid (x(.), u(.)) \text{ solution of 2.1}, x(t) = \bar{x}\}$$

where $Q_\Omega := \{(t,x) \mid t \in \mathbb{R}_0^+,\ x \in \Omega(t)\} \subset \mathbb{R}_0^+ \times \mathbb{R}^n$. Analogously, we denote by $V^\star : Q_\Omega \to [-\infty, +\infty]$ the value function of the problem ($\mathcal{P}^\star$).

Next, we state the main result of this section

Theorem 2.5.4 (Lipschitz continuity) *Consider Assumptions 2.5.1 and 2.3.2. Suppose that* (2.21) *hold true and* $\limsup_{t\to+\infty} \frac{1}{t}\int_0^t k(\tau)\,d\tau < +\infty$. *Then there exist* $b > 1$ *and* $K > 0$ *such that for all* $\lambda > K$ *we have:*

(i) $V^\star(t,.)$ *is* $b \cdot e^{-(\lambda-K)t}$*-Lipschitz continuous on* $\Omega(t)$*, for any* $t \geq 0$*;*
(ii) $\lim_{t\to+\infty} V^\star(t, x(t)) = 0$ *for any feasible trajectory* $x(.)$*;*
(iii) $V^\star = V$ *on* Q_Ω.

Moreover, the same conclusions hold for the autonomous case replacing the above assumptions with Assumptions 2.3.4, 2.3.5.

We show first the following.

Proposition 2.5.5 *Consider Assumptions 2.3.1 and 2.3.2. Suppose that*

$$\limsup_{t\to+\infty} \frac{1}{t}\int_0^t \varphi(\tau)\,d\tau < +\infty.$$

Then there exist $C > 1$ *and* $K > 0$ *such that for any* $t_0 \geq 0$*, any* $x^0, x^1 \in \Omega(t_0)$*, and any feasible* F_∞*-trajectory* $x : \mathbb{R}_{t_0}^+ \to \mathbb{R}^n$*, with* $x(t_0) = x^0$*, we can find a feasible* F_∞*-trajectory* $\tilde{x} : \mathbb{R}_{t_0}^+ \to \mathbb{R}^n$*, with* $\tilde{x}(t_0) = x^1$*, such that*

$$|\tilde{x}(t) - x(t)| \le Ce^{Kt}|x^1 - x^0| \quad \forall t \in \mathbb{R}^+_{t_0}.$$

Moreover, the same conclusions hold for the autonomous case replacing the above assumptions with Assumptions 2.3.4 and 2.3.5.

Proof Let $\delta = 1$ and $\beta > 0$ be as in Theorem 2.3.3. Consider $K_1 > 0$, $K_2 > 0$, and $\tilde{k} > 0$ such that

$$2\beta + 1 < e^{K_1} \quad \text{and} \quad \int_0^{t+1} \varphi(s)\,ds \le K_2 t + \tilde{k} \quad \forall t \in \mathbb{R}^+_0. \tag{2.50}$$

Fix $t_0 \ge 0, x^0, x^1 \in \Omega(t_0)$, with $x^1 \neq x^0$, and a feasible F_∞-trajectory $x : \mathbb{R}^+_{t_0} \to \mathbb{R}^n$ with $x(t_0) = x_0$. By the Generalized Filippov Existence Theorem, there exists an F-trajectory $y_0 : [t_0, t_0 + 1] \to \mathbb{R}^n$ such that $y_0(t_0) = x^1$ and

$$\|y_0 - x\|_{\infty,[t_0,t_0+1]} \le e^{\int_{t_0}^{t_0+1} \varphi(s)\,ds}|x^1 - x^0|.$$

Denote by $x_0 : [t_0, t_0 + 1] \to \mathbb{R}^n$ the feasible F-trajectory, with $x_0(t_0) = x^1$, satisfying the conclusions of Theorem 2.3.3 with $\hat{x}(.) = y_0(.)$. Thus

$$\begin{aligned}\|x_0 - y_0\|_{\infty,[t_0,t_0+1]} &\le \beta\left(\max_{t\in[t_0,t_0+1]} d_{\Omega(t)}(y_0(t)) + |x^1 - x^0|\right)\\ &\le \beta(\|y_0 - x\|_{\infty,[t_0,t_0+1]} + |x^1 - x^0|)\\ &\le 2\beta e^{\int_{t_0}^{t_0+1} \varphi(s)\,ds}|x^1 - x^0|,\end{aligned}$$

and therefore

$$\begin{aligned}\|x_0 - x\|_{\infty,[t_0,t_0+1]} &\le \|x_0 - y_0\|_{\infty,[t_0,t_0+1]} + \|y_0 - x\|_{\infty,[t_0,t_0+1]}\\ &\le (2\beta + 1)e^{\int_{t_0}^{t_0+1} \varphi(s)\,ds}|x^1 - x^0|.\end{aligned} \tag{2.51}$$

Now, applying again the Generalized Filippov Existence Theorem on $[t_0 + 1, t_0 + 2]$, there exists an F-trajectory $y_1 : [t_0 + 1, t_0 + 2] \to \mathbb{R}^n$, with $y_1(t_0 + 1) = x_0(t_0 + 1)$, such that, thanks to (2.51),

$$\|y_1 - x\|_{\infty,[t_0+1,t_0+2]} \le (2\beta + 1)e^{\int_{t_0}^{t_0+2} \varphi(s)\,ds}|x^1 - x^0|. \tag{2.52}$$

Denoting by $x_1 : [t_0 + 1, t_0 + 2] \to \mathbb{R}^n$ the feasible F-trajectory, with $x_1(t_0 + 1) = x_0(t_0 + 1)$, satisfying the conclusions of Theorem 2.3.3, for $\hat{x}(.) = y_1(.)$, we deduce from (2.52), that

$$\|x_1 - y_1\|_{\infty,[t_0+1,t_0+2]} \le \beta(2\beta + 1)e^{\int_{t_0}^{t_0+2} \varphi(s)\,ds}|x^1 - x^0|. \tag{2.53}$$

Hence, taking note of (2.52) and (2.53),

$$\|x_1 - x\|_{\infty,[t_0+1,t_0+2]} \leq (2\beta+1)^2 e^{\int_{t_0}^{t_0+2} \varphi(s)\,ds} |x^1 - x^0|.$$

Continuing this construction, we obtain a sequence of feasible F-trajectories $x_i : [t_0+i, t_0+i+1] \to \mathbb{R}^n$ such that $x_j(t_0+j) = x_{j-1}(t_0+j)$ for all $j \geq 1$, and

$$\|x_i - x\|_{\infty,[t_0+i,t_0+i+1]} \leq (2\beta+1)^{i+1} e^{\int_{t_0}^{t_0+i+1} \varphi(s)\,ds} |x^1 - x^0| \;\; \forall i \in \mathbb{N}. \tag{2.54}$$

Define the feasible F_∞-trajectory $\tilde{x} : \mathbb{R}^+_{t_0} \to \mathbb{R}^n$ by $\tilde{x}(t) := x_i(t)$ if $t \in [t_0+i, t_0+i+1]$ and observe that $\tilde{x}(t_0) = x^1$. Let $t \geq t_0$. Then there exists $i \in \mathbb{N}$ such that $t \in [t_0+i, t_0+i+1]$. So, from (2.54) and (2.50), it follows that

$$\begin{aligned}
|\tilde{x}(t) - x(t)| &\leq (2\beta+1)^{i+1} e^{\int_{t_0}^{t_0+i+1} \varphi(s)\,ds} |x^1 - x^0| \\
&\leq e^{\tilde{k}}(2\beta+1) e^{(K_1+K_2)(t_0+i)} |x^1 - x^0| \\
&\leq C e^{Kt} |x^1 - x^0|,
\end{aligned}$$

where $K = K_1 + K_2$ and $C = e^{\tilde{k}}(2\beta+1)$.

Following the same arguments and using Theorem 2.3.6, all the above conclusions follows for the autonomous case. □

Proof of Theorem 2.5.4 We notice that, by Theorem 2.2.4, the problem $(\mathcal{P}^\star)$ admits feasible trajectory-control pairs for any initial condition; using the sub-linear growth of f and Gronwall's Lemma, we have $1 + |x(t)| \leq (1 + |x_0|) e^{\int_{t_0}^t c(s)\,ds}$ for all $t \geq t_0$ and for any trajectory-control pair $(x(.), u(.))$ at $t_0 \in \mathbb{R}^+_0$, $x_0 \in \Omega(t_0)$.

Next, we show *(i)*. Let $a_1 > 0$, $a_2 > 0$ be such that

$$\int_0^t c(s)\,ds \leq a_1 t + a_2 \qquad \forall t \in \mathbb{R}^+_0. \tag{2.55}$$

For all $T > t_0$, we have

$$\begin{aligned}
\int_{t_0}^T e^{-\lambda t} |L^\star(t, x(t), w(t))|\,dt &\leq (n+1) \int_{t_0}^T e^{-\lambda t} c(t)(1 + |x_0|) e^{\int_{t_0}^t c(s)\,ds}\,dt \\
&\leq (n+1)(1+|x_0|) e^{a_2} \int_{t_0}^T e^{-(\lambda - a_1)t} c(t)\,dt.
\end{aligned} \tag{2.56}$$

Then, by (2.55) and denoting $\psi(t) = \int_{t_0}^t c(s)\,ds$, for any $\lambda > a_1$

$$\begin{aligned}
&\int_{t_0}^T e^{-\lambda t} |L^\star(t, x(t), w(t))|\,dt \\
&\leq (n+1)(1+|x_0|) e^{a_2} \left(\left[e^{-(\lambda-a_1)t} \psi(t) \right]_{t_0}^T + (\lambda - a_1) \int_{t_0}^T e^{-(\lambda-a_1)t} \psi(t)\,dt \right) \\
&\leq (n+1)(1+|x_0|) e^{a_2} \left(e^{-(\lambda-a_1)T} (a_1 T + a_2) + \left(a_1 t_0 + \tfrac{a_1}{\lambda - a_1} + a_2 \right) e^{-(\lambda - a_1)t_0} \right).
\end{aligned} \tag{2.57}$$

Passing to the limit when $t \to +\infty$, we deduce that for every feasible trajectory-control pair $(x(.), w(.))$ at (t_0, x_0)

$$\int_{t_0}^{+\infty} e^{-\lambda t}|L^\star(t, x(t), w(t))|\,dt < +\infty \qquad \forall \lambda > a_1.$$

From now on, assume that $\lambda > a_1$. Fix $t \geq 0$ and $x^1, x^0 \in \Omega(t)$ with $x^1 \neq x^0$. Then, for any $\delta > 0$ there exists a feasible trajectory-control pair $(x_\delta(.), w_\delta(.))$ at (t, x^0) such that

$$V^\star(t, x^0) + e^{-\delta t}|x^1 - x^0| > \int_t^{+\infty} e^{-\lambda s} L^\star(s, x_\delta(s), w_\delta(s))\,ds.$$

Hence

$$\begin{aligned} V^\star(t, x^1) - V^\star(t, x^0) &\leq e^{-\delta t}|x^1 - x^0| \\ &+ \lim_{\tau\to+\infty} \left| \int_t^\tau e^{-\lambda s} L^\star(s, x(s), w(s))\,ds - \int_t^\tau e^{-\lambda s} L^\star(s, x_\delta(s), w_\delta(s))\,ds \right| \end{aligned} \tag{2.58}$$

for any feasible trajectory-control pair $(x(.), w(.))$ satisfying $x(t) = x^1$. Consider the following state constrained differential inclusion in $\mathbb{R}^{n+1}$

$$\begin{cases} (x, z)'(s) \in F^\star(s, x(s), z(s)) & \text{a.e. } s \in \mathbb{R}_t^+, \\ x(s) \in \Omega(s) & \forall s \in \mathbb{R}_t^+, \end{cases}$$

where for all $(t, x, z) \in \mathbb{R}_0^+ \times \mathbb{R}^n \times \mathbb{R}$ the time-measurable set-valued maps

$$F^\star(t, x, z) := \{(f^\star(t, x, w), L^\star(t, x, w)) \mid w \in W(t)\}.$$

Putting $z_\delta(s) = \int_t^s L^\star(\xi, x_\delta(\xi), w_\delta(\xi))\,d\xi$, applying Proposition 2.5.5 with $F = F^\star$ and the Measurable Selection Theorem, there exist $C > 1$ and $K > 0$ such that for all $\delta > 0$ we can find a $F^\star_\infty$-trajectory $(\tilde{x}_\delta(.), \tilde{z}_\delta(.))$ on $\mathbb{R}_t^+$, and a measurable selection $\tilde{w}_\delta(s) \in W(s)$ a.e. $s \geq t$, satisfying

$$\begin{cases} (\tilde{x}_\delta, \tilde{z}_\delta)'(s) = (f^\star(s, \tilde{x}_\delta(s), \tilde{w}_\delta(s)), L^\star(s, \tilde{x}_\delta(s), \tilde{w}_\delta(s))) & \text{a.e. } s \in \mathbb{R}_t^+, \\ (\tilde{x}_\delta(t), \tilde{z}_\delta(t)) = (x^1, 0), & \\ \tilde{x}_\delta(s) \in \Omega(s) & \forall s \in \mathbb{R}_t^+, \end{cases}$$

and for any $s \geq t$

$$|\tilde{x}_\delta(s) - x_\delta(s)| + |\tilde{z}_\delta(s) - z_\delta(s)| \leq C e^{Ks}|x^1 - x^0|. \tag{2.59}$$

Now, relabeling by K the constant $K \vee a_1$, by (2.59) and integrating by parts, for all $\lambda > K$, all $\tau \geq t$, and all $\delta > 0$

$$
\begin{aligned}
&\left|\int_t^{\tau} e^{-\lambda s} L^{\star}(s, \tilde{x}_\delta(s), \tilde{w}_\delta(s))\, ds - \int_t^{\tau} e^{-\lambda s} L^{\star}(s, x_\delta(s), w_\delta(s))\, ds\right| \\
&\le \left|\left[e^{-\lambda s}\left(\int_t^{s} L^{\star}(\xi, \tilde{x}_\delta(\xi), \tilde{w}_\delta(\xi))\, d\xi - \int_t^{s} L^{\star}(\xi, x_\delta(\xi), w_\delta(\xi))\, d\xi\right)\right]_t^{\tau}\right| \\
&\quad + \lambda\left|\int_t^{\tau} e^{-\lambda s}\left(\int_t^{s} L^{\star}(\xi, \tilde{x}_\delta(\xi), \tilde{w}_\delta(\xi))\, d\xi - \int_t^{s} L^{\star}(\xi, x_\delta(\xi), w_\delta(\xi))\, d\xi\right) ds\right| \\
&\le e^{-\lambda\tau}|\tilde{z}_\delta(\tau) - z_\delta(\tau)| + \lambda\int_t^{\tau} e^{-\lambda s}|\tilde{z}_\delta(s) - z_\delta(s)| ds \\
&\le Ce^{-\lambda\tau}e^{K\tau}|x^1 - x^0| + \lambda C\int_t^{\tau} e^{-(\lambda-K)s}|x^1 - x^0|\, ds \\
&= \left(Ce^{-(\lambda-K)\tau} + \lambda C\left[-\frac{e^{-(\lambda-K)s}}{\lambda-K}\right]_t^{\tau}\right)|x^1 - x^0| \\
&= \left(-\frac{CK}{\lambda-K}e^{-(\lambda-K)\tau} + \frac{\lambda C}{\lambda-K}e^{-(\lambda-K)t}\right)|x^1 - x^0| \\
&\le \frac{\lambda C}{\lambda-K}e^{-(\lambda-K)t}|x^1 - x^0|.
\end{aligned}
\tag{2.60}
$$

Taking note of (2.58), (2.60), and putting $\delta = \lambda - K$, for all $\lambda > K$ we get

$$
V^{\star}(t, x^1) - V^{\star}(t, x^0) \le \left(\frac{\lambda C}{\lambda - K} + 1\right) e^{-(\lambda-K)t}|x^1 - x^0|.
$$

By the symmetry of the previous inequality with respect to x^1 and x^0, and since λ, C, and K do not depend on t, x^1, and x^0, the statement *(i)* follows.

Now, let $(t_0, x_0) \in Q_\Omega$ and consider a feasible trajectory $X(.)$ at (t_0, x_0). Let $t > t_0$ and $(x(.), w(.))$ be a feasible trajectory-control pair at $(t, X(t))$ such that $V^{\star}(t, X(t)) > \int_t^{+\infty} e^{-\lambda s} L^{\star}(s, x(s), w(s))\, ds - \frac{1}{t}$. Then

$$
|V^{\star}(t, X(t))| \le \int_t^{+\infty} e^{-\lambda s}|L^{\star}(s, x(s), w(s))|\, ds + \frac{1}{t}.
$$

From (2.55) and (2.56), we have for all $T > t$

$$
\begin{aligned}
&\int_t^{T} e^{-\lambda s}|L^{\star}(s, x(s), w(s))|\, ds \le \int_t^{T} e^{-\lambda s}(1 + |X(t)|)e^{\int_t^s c(\omega)\, d\omega}c(s)\, ds. \\
&\le (1 + |x_0|)\int_t^{T} e^{-\lambda s}e^{\int_{t_0}^t c(\omega)\, d\omega}e^{\int_t^s c(\omega)\, d\omega}c(s)\, ds \\
&\le (1 + |x_0|)\int_t^{T} e^{-\lambda s}e^{\int_0^s c(\omega)\, d\omega}c(s)\, ds \\
&\le (1 + |x_0|)e^{a_2}\int_t^{T} e^{-(\lambda-a_1)s}c(s)\, ds.
\end{aligned}
$$

Then, arguing as in (2.57) with t_0 replaced by t and taking the limit when $t \to +\infty$, we deduce that

$$|V^\star(t, X(t))| \leq (1 + |x_0|)e^{a_2} \left(a_1 t + \frac{a_1}{\lambda - a_1} + a_2 \right) e^{-(\lambda - a_1)t} + \frac{1}{t}.$$

Since $K \geq a_1$, *(ii)* follows passing to the limit when $t \to +\infty$.

Next, we show *(iii)*. Notice that $V^\star(t, x) \leq V(t, x)$ for any $(t, x) \in Q_\Omega$, and $V^\star(t, .)$ is Lipschitz continuous on $\Omega(t)$ for all $t \geq 0$ whenever $\lambda > 0$ is sufficiently large. Fix $t_0 \in \mathbb{R}_0^+$, $x_0 \in \Omega(t_0)$, and $\varepsilon > 0$. We claim that: for all $j \in \mathbb{N}^+$ there exists a finite set of trajectory-control pairs $\{(x_k(.), u_k(.))\}_{k=1,\dots,j}$ satisfying the following: $x_k'(s) = f(s, x_k(s), u_k(s))$ a.e. $s \in [t_0, t_0 + k]$ and $x_k(s) \in \Omega(s)$ for all $s \in [t_0, t_0 + k]$ and for all $k = 1, \dots, j$; if $j \geq 2$, $x_k|_{[t_0, t_0+k-1]}(.) = x_{k-1}(.)$ for all $k = 2, \dots, j$; and for all $k = 1, \dots, j$

$$V^\star(t_0, x_0) \geq V^\star(t_0 + k, x_k(t_0 + k)) + \int_{t_0}^{t_0+k} e^{-\lambda t} L(t, x_k(t), u_k(t))\, dt - \varepsilon \sum_{i=1}^{k} \frac{1}{2^i}. \tag{2.61}$$

We prove the claim by the induction argument with respect to $j \in \mathbb{N}^+$. By the dynamic programming principle, there exists a trajectory-control pair $(\tilde{x}(.), \tilde{w}(.))$ on $[t_0, t_0 + 1]$, feasible for the problem $(\mathcal{P}^\star)$ at (t_0, x_0), such that

$$V^\star(t_0, x_0) + \frac{\varepsilon}{4} > V^\star(t_0 + 1, \tilde{x}(t_0 + 1)) + \int_{t_0}^{t_0+1} e^{-\lambda t} L^\star(t, \tilde{x}(t), \tilde{w}(t))\, dt. \tag{2.62}$$

By Lemma 1.3.3, for any $h > 0$ there exists a measurable control $\hat{u}^h(t) \in U(t)$ a.e. $t \in [t_0, t_0 + 1]$ such that the solution of the equation $(\hat{x}^h)'(t) = f(t, \hat{x}^h(t), \hat{u}^h(t))$ a.e. $t \in [t_0, t_0 + 1]$, with $\hat{x}^h(t_0) = x_0$, satisfies

$$\|\hat{x}^h - \tilde{x}\|_{\infty, [t_0, t_0+1]} < h$$

and

$$\left| \int_{t_0}^{t_0+1} e^{-\lambda t} L^\star(t, \tilde{x}(t), \tilde{w}(t))\, dt - \int_{t_0}^{t_0+1} e^{-\lambda t} L(t, \hat{x}^h(t), \hat{u}^h(t))\, dt \right| < h.$$

Now, consider the following state constrained differential inclusion in $\mathbb{R}^{n+1}$

$$\begin{cases} (x, z)'(s) \in F(s, x(s), z(s)) & \text{a.e. } s \in [t_0, t_0 + 1], \\ x(s) \in \Omega(s) & \forall s \in [t_0, t_0 + 1], \end{cases}$$

where

$$F(t, x, z) := \{(f(t, x, u), e^{-\lambda t} L(t, x, u)) \mid u \in U(t)\}.$$

Letting $\hat{X}^h(.) = (\hat{x}^h(.), \hat{z}^h(.))$, with $\hat{z}^h(t) = \int_{t_0}^t e^{-\lambda s} L(s, \hat{x}^h(s), \hat{u}^h(s))\, ds$, by Theorem 2.3.3 and the Measurable Selection Theorem, there exist $\beta > 0$ (not depending on (t_0, x_0)) such that for any $h > 0$ we can find a feasible F-trajectory $X^h(.) = (x^h(.), z^h(.))$ on $[t_0, t_0 + 1]$, with $X^h(t_0) = (x_0, 0)$, and a measurable control $u^h(s) \in U(s)$ for a.e. $s \in [t_0, t_0 + 1]$, such that for $s \in [t_0, t_0 + 1]$

$$(x^h, z^h)'(s) = (f(s, x^h(s), u^h(s)), e^{-\lambda s} L(s, x^h(s), u^h(s)))$$

and

$$\|X^h - \hat{X}^h\|_{\infty,[t_0,t_0+1]} \le \beta \cdot \left(\sup_{s\in[t_0,t_0+1]} d_{\Omega(s)\times\mathbb{R}}(\hat{X}^h(s)) + h \right).$$

Since $\sup_{s\in[t_0,t_0+1]} d_{\Omega(s)\times\mathbb{R}}(\hat{X}^h(s)) \le \|\tilde{x} - \hat{x}^h\|_{\infty,[t_0,t_0+1]}$, we have

$$\begin{aligned}
&\left| \int_{t_0}^{t_0+1} e^{-\lambda t} L(t, x^h(t), u^h(t))\, dt - \int_{t_0}^{t_0+1} e^{-\lambda t} L^\star(t, \tilde{x}(t), \tilde{w}(t))\, dt \right| \\
&\le \left| \int_{t_0}^{t_0+1} e^{-\lambda t} L^\star(t, \tilde{x}(t), \tilde{w}(t))\, dt - \int_{t_0}^{t_0+1} e^{-\lambda t} L(t, \hat{x}^h(t), \hat{u}^h(t))\, dt \right| \\
&\quad + \left| \int_{t_0}^{t_0+1} e^{-\lambda t} L(t, x^h(t), u^h(t))\, dt - \int_{t_0}^{t_0+1} e^{-\lambda t} L(t, \hat{x}^h(t), \hat{u}^h(t))\, dt \right| \\
&< h(2\beta + 1)
\end{aligned}$$

and

$$\begin{aligned}
\|x^h - \tilde{x}\|_{\infty,[t_0,t_0+1]} &\le \|\tilde{x} - \hat{x}^h\|_{\infty,[t_0,t_0+1]} + \|x^h - \hat{x}^h\|_{\infty,[t_0,t_0+1]} \\
&< h(2\beta + 1).
\end{aligned}$$

Hence, choosing $0 < h < \varepsilon/4(2\beta + 1)$ sufficiently small, we can find a trajectory-control pair $(x^h(.), u^h(.))$ on $[t_0, t_0 + 1]$, with $u^h(s) \in U(s)$ and $(x^h)'(s) = f(s, x^h(s), u^h(s))$ a.e. $s \in [t_0, t_0 + 1]$, $x^h(t_0) = x_0$, and $x^h(s) \in \Omega(s)$ for $s \in [t_0, t_0 + 1]$, such that, by (2.62) and continuity of $V^\star(t_0 + 1, .)$

$$V^\star(t_0, x_0) > V^\star(t_0 + 1, x^h(t_0 + 1)) + \int_{t_0}^{t_0+1} e^{-\lambda t} L(t, x^h(t), u^h(t))\, dt - \frac{\varepsilon}{2}.$$

Letting $(x_1(.), u_1(.)) := (x^h(.), u^h(.))$, the conclusion follows for $j = 1$. Now, suppose we have shown that there exist $\{(x_k(.), u_k(.))\}_{k=1,\dots,j}$ satisfying the claim. Let us to prove it for $j + 1$. By the dynamic programming principle there exists a trajectory-control pair $(\tilde{x}(.), \tilde{w}(.))$ on $[t_0 + j, t_0 + j + 1]$, feasible for the problem $(\mathcal{P}^\star)$ at $(t_0 + j, x_j(t_0 + j))$, such that

$$\begin{aligned}
V^\star(t_0 + j, x_j(t_0 + j)) + \tfrac{\varepsilon}{2^{j+2}} > {} & V^\star(t_0 + j + 1, \tilde{x}(t_0 + j + 1)) \\
& + \int_{t_0+j}^{t_0+j+1} e^{-\lambda t} L^\star(t, \tilde{x}(t), \tilde{w}(t))\, dt.
\end{aligned} \tag{2.63}$$

As before, for every $h > 0$ there exist a feasible F-trajectory $X^h(.) = (x^h(.), z^h(.))$ on $[t_0 + j, t_0 + j + 1]$, with $X^h(t_0) = (x_j(t_0 + j), 0)$, and a measurable control $u^h(s) \in U(s)$ a.e. $s \in [t_0 + j, t_0 + j + 1]$, such that for a.e. $s \in [t_0 + j, t_0 + j + 1]$

$$(x^h, z^h)'(s) = (f(s, x^h(s), u^h(s)), e^{-\lambda s} L(s, x^h(s), u^h(s))),$$

$$\left| \int_{t_0+j}^{t_0+j+1} e^{-\lambda t} L(t, x^h(t), u^h(t))\, dt - \int_{t_0+j}^{t_0+j+1} e^{-\lambda t} L^\star(t, \tilde{x}(t), \tilde{w}(t))\, dt \right| < h(2\beta + 1),$$

and

$$\|x^h - \tilde{x}\|_{\infty,[t_0+j,t_0+j+1]} < h(2\beta + 1).$$

Putting

$$(x_{j+1}(.), u_{j+1}(.)) := \begin{cases} (x_j(.), u_j(.)) \text{ on } [t_0, t_0 + j], \\ (x^h(.), u^h(.)) \text{ on } [t_0 + j, t_0 + j + 1], \end{cases} \tag{2.64}$$

and choosing $0 < h < \varepsilon/2^{j+2}(2\beta + 1)$ sufficiently small, it follows from (2.63) that

$$\begin{aligned} V^\star(t_0 + j, x_j(t_0 + j)) \geq V^\star&(t_0 + j + 1, x_{j+1}(t_0 + j + 1)) \\ &+ \textstyle\int_{t_0+j}^{t_0+j+1} e^{-\lambda t} L(t, x_{j+1}(t), u_{j+1}(t))\, dt - \frac{2\varepsilon}{2^{j+2}}. \end{aligned} \tag{2.65}$$

So, taking note of (2.64) and (2.65), we obtain

$$\begin{aligned} &V^\star(t_0, x_0) \\ &\geq V^\star(t_0 + j, x_j(t_0 + j)) + \int_{t_0}^{t_0+j} e^{-\lambda t} L(t, x_j(t), u_j(t))\, dt - \varepsilon \sum_{i=1}^{j} \frac{1}{2^i} \\ &\geq V^\star(t_0 + j + 1, x_{j+1}(t_0 + j + 1)) - \varepsilon \sum_{i=1}^{j} \frac{1}{2^i} - \frac{\varepsilon}{2^{j+1}} \\ &\quad + \int_{t_0+j}^{t_0+j+1} e^{-\lambda t} L(t, x_{j+1}(t), u_{j+1}(t))\, dt \\ &\quad + \int_{t_0}^{t_0+j} e^{-\lambda t} L(t, x_j(t), u_j(t))\, dt \\ &= V^\star(t_0 + j + 1, x_{j+1}(t_0 + j + 1)) \\ &\quad + \int_{t_0}^{t_0+j+1} e^{-\lambda t} L(t, x_{j+1}(t), u_{j+1}(t))\, dt - \varepsilon \sum_{i=1}^{j+1} \frac{1}{2^i}. \end{aligned}$$

Hence $\{(x_k(.), u_k(.))\}_{k=1,\ldots,j+1}$ also satisfy our claim. Now, let us define the trajectory-control pair $(x(.), u(.))$ by $(x(t), u(t)) := (x_k(t), u_k(t))$ if $t \in [t_0 + k -$

$1, t_0 + k]$. Then $(x(.), u(.))$ is a feasible trajectory-control pair for the problem $(\mathcal{P})$ at (t_0, x_0). Since, thanks to *(ii)*, $V^\star(t, x(t)) \to 0$ when $t \to +\infty$, from (2.61) we have

$$V^\star(t_0, x_0) \geq \int_{t_0}^{+\infty} e^{-\lambda t} L(t, x(t), u(t))\, dt - \varepsilon.$$

Hence, we deduce that (t_0, x_0) lays in the domain of the value function V and so $V^\star(t_0, x_0) \geq V(t_0, x_0) - \varepsilon$. From the arbitrariness of ε, the conclusion follows.

Following the same arguments and using Theorem 2.3.6 and Proposition 2.5.5, all the above conclusions hold for the autonomous case. □

Corollary 2.5.6 *Consider any* $N > 0$ *with*

$$N \geq \sup\{|f(t, x, u)| + |L(t, x, u)| \mid t \geq 0,\ x \in \mathbb{R}^n,\ u \in U(t)\} < +\infty.$$

Then, for any $\lambda > 0$ *sufficiently large, for any* $t \geq 0$*, and any* $x \in \Omega(t)$*, the function* $V(., x)$ *is Lipschitz continuous on* $\mathbb{R}_t^+$ *with constant* $\left(L(t) + 2e^{-\lambda t}\right) N$ *and* $L(t) := be^{-(\lambda - K)t}$.

Proof From Theorem 2.5.4, when $\lambda > 0$ is large enough, $V(t, .)$ is $L(t)$-Lipschitz continuous on $\Omega(t)$. Fix $t \geq 0$ and $x \in \Omega(t)$. Let $s, \tilde{s} \in \mathbb{R}_t^+$.

Suppose that $s \geq \tilde{s}$. Then, by the dynamic programming principle, there exists a feasible trajectory-control pair $(\bar{x}(.), \bar{u}(.))$ at $(\tilde{s}, x)$ such that

$$\begin{aligned} V(s, x) - V(\tilde{s}, x) &\leq |V(s, x) - V(s, \bar{x}(s))| + \int_{\tilde{s}}^{s} e^{-\lambda \xi} |L(\xi, \bar{x}(\xi), \bar{u}(\xi))|\, d\xi \\ &\quad + N|s - \tilde{s}|e^{-\lambda t} \\ &\leq L(s)N|s - \tilde{s}| + N|s - \tilde{s}|e^{-\lambda \tilde{s}} + N|s - \tilde{s}|e^{-\lambda t} \\ &\leq \left(L(t) + 2e^{-\lambda t}\right) N|s - \tilde{s}|. \end{aligned} \tag{2.66}$$

Arguing in a similar way, we get (2.66) when $s < \tilde{s}$. Hence, by the symmetry with respect to s and $\tilde{s}$ in (2.66), the conclusion follows. □

2.5.1 *The Autonomous Case*

The autonomous case, where dynamics and costs do not explicitly depend on time, exhibits particular regularity properties. This subsection focuses on uniform local Lipschitz continuity of the class of value functions for autonomous systems, exploiting time invariance to obtain stronger results compared to the general case.

In this section, we assume that:

- f is time independent, i.e., $f(t, x, u) = f(x, u)$;
- $U(.) \equiv U$ is closed;
- $\Omega(.) \equiv \Omega$.

Theorem 2.5.7 *Assume that:*

- $L(t,x,u) = e^{-\lambda t} l(x,u)$;
- *Assumption 2.5.1 hold true with Assumption 2.5.1:(IV.2) replaced by*

$$|f(x,u)| + |l(x,u)| \leq c(t)(1+|x|) \quad \forall u \in U,$$

where $c(t) = K(1+e^{-\lambda t})$ *and* $K > 0$.

Suppose that dom $v \neq \emptyset$ *and* v *is locally bounded on it. Then the function* $v(.) := V(0,.)$ *is locally Lipschitz continuous on its domain for all large* $\lambda > 0$.

Consequently, the value function $V(t,x)$ *of problem* $\mathcal{B}_\infty$ *is locally Lipschitz continuous on its domain uniformly in* $t \in \mathbb{R}_0^+$ *for all large* $\lambda > 0$.

Remark 2.5.8 We would like to notice that, under Assumption 2.5.1, if A is compact then dom $v = A$ and v is bounded on it.

We need first the following technical result.

Lemma 2.5.9 *Let* $R > 0$ *and* $\omega : [0, R] \to \mathbb{R}_0^+$ *be a nondecreasing function. Then the following statements hold true:*

- *If there exist* $\alpha > 0,\ \beta \geq 1$ *such that*

$$\omega(r) \leq \alpha r + \omega(\beta r) \qquad \forall 0 \leq r \leq R/\beta, \tag{2.67}$$

then there exists a constant $C_R \geq 0$ *such that*

$$\omega(r) \leq C_R r \qquad \forall\, 0 \leq r \leq R.$$

- *If there exist* $0 < \theta < 1,\ \alpha > 0,\ \beta \geq 1$ *such that*

$$\omega(r) \leq \alpha r + \theta\omega(\beta r) \qquad \forall 0 \leq r \leq R/\beta, \tag{2.68}$$

then, letting any $m \geq 1$ *be a real number such that* $\theta^m \beta < 1$*, there exists a constant* $C_R \geq 0$ *such that*

$$\omega(r) \leq C_R {r}^{1/m} \qquad \forall\, 0 \leq r \leq R.$$

Proof We prove the first statement. Let $0 < \tau < 1$ be such that $\tau\beta \leq 1$. Then $\tau R \leq \dfrac{R}{\beta}$ and by the growth assumption in (2.67) and the monotonicity of ω, we have that

$$\begin{aligned} \omega(\tau R) &\leq \alpha\tau R + \omega(\beta\tau R) \\ &\leq \alpha\tau R + \omega(R). \end{aligned} \tag{2.69}$$

Applying again (2.68) and (2.69) we obtain

$$\begin{aligned}\omega(\tau^2 R) &\le \alpha\tau^2 R + \omega(\tau R)\\ &\le \alpha\tau^2 R + (\alpha\tau R + \omega(R))\\ &= \alpha\tau R(\tau+1) + \omega(R).\end{aligned}$$

So, by induction on $k \in \mathbb{N}$ it is straightforward to show that

$$\begin{aligned}\omega(\tau^k R) &\le \alpha\tau R(\tau^{k-1} + \tau^{k-2} + \cdots + 1) + \omega(R)\\ &= \alpha R\tau^k\left(1 + \frac{1}{\tau} + \cdots + \left(\frac{1}{\tau}\right)^k\right) + \omega(R)\\ &< \alpha R\tau^k \frac{1}{1 - 1/\tau} + \omega(R)\\ &= \frac{\alpha R}{\tau - 1}\tau^{k+1} + \omega(R).\end{aligned}$$

Now let $r \in [0, R]$. Then there exists $k \in \mathbb{N}$ such that $\tau^{k+1}R < r \le \tau^k R$. Finally

$$\begin{aligned}\omega(r) &\le \frac{\alpha R}{\tau - 1}\tau^{k+1} + \omega(R)\\ &\le \frac{\alpha}{\tau - 1}\tau^{k+1}R + \tau^{k+1}R\frac{\omega(R)}{\tau R}\\ &\le \left(\frac{\alpha}{\tau - 1} + \frac{\omega(R)}{\tau R}\right) r.\end{aligned}$$

So, the first statement holds true with $C = \dfrac{\alpha}{\tau - 1} + \dfrac{\omega(R)}{\tau R}$.

Next, we show the second statement.

Suppose first that $m = 1$. Let $\theta < \tau < 1$ be such that $\tau\beta \le 1$. Using the same iterative argument as in the proof of the first statement, we can conclude that for every $k \in \mathbb{N}$

$$\begin{aligned}\omega(\tau^k R) &\le \alpha\tau R(\tau^{k-1} + \theta\tau^{k-2} + \cdots + \theta^{k-1}) + \theta^k\omega(R)\\ &< \frac{\alpha R}{\tau - \theta}\tau^{k+1} + \theta^k\omega(R).\end{aligned}$$

Again, let $r \in [0, R]$ and keep $k \in \mathbb{N}$ such that $\tau^{k+1}R < r \le \tau^k R$. Hence

$$\begin{aligned}\omega(r) &\le \frac{\alpha R}{\tau - \theta}\tau^{k+1} + \theta^k\omega(R)\\ &\le \frac{\alpha}{\tau - \theta}\tau^{k+1}R + \tau^{k+1}R\frac{\omega(R)}{\tau R}\\ &\le \left(\frac{\alpha}{\tau - \theta} + \frac{\omega(R)}{\tau R}\right) r.\end{aligned}$$

The conclusion holds true with $C = \dfrac{\alpha}{\tau - \theta} + \dfrac{\omega(R)}{\tau R}$.

If $m > 1$, by the growth assumption in (2.68) and the monotonicity of ω we have that

$$\begin{aligned}\omega(\theta^m R) &\leq \alpha\theta^m R + \theta\omega(\beta\theta^m R) \\ &\leq \alpha\theta^m R + \theta\omega(R).\end{aligned} \tag{2.70}$$

Applying again (2.68), monotonicity, and (2.70) we obtain

$$\begin{aligned}\omega(\theta^{2m} R) &\leq \alpha\theta^{2m} R + \theta\omega(\theta^m R) \\ &\leq \alpha\theta^{2m} R + \theta\left(\alpha\theta^m R + \theta\omega(R)\right) \\ &= \alpha\theta^{m+1} R(1 + \theta^{m-1}) + \theta^2\omega(R).\end{aligned}$$

So, by induction on $k \in \mathbb{N}$ we have that

$$\begin{aligned}\omega(\theta^{km} R) &\leq \alpha\theta^{m+k-1} R(1 + \theta^{m-1} + \cdots + \theta^{(k-1)(m-1)}) + \theta^k\omega(R) \\ &< \alpha R\theta^{m+k-1}\frac{1}{1 - \theta^{m-1}} + \theta^k\omega(R) \\ &= \left(\frac{\alpha R\theta^{m-1}}{1 - \theta^{m-1}} + \omega(R)\right)\theta^k.\end{aligned}$$

Let $r \in [0, R]$. Then there exists $k \in \mathbb{N}$ such that $\theta^{(k+1)m} R < r \leq \theta^{km} R$. Thus,

$$\omega(r) \leq \omega(\theta^{km} R) \leq \tilde{C}\theta^k < \frac{\tilde{C}}{\theta}\left(\frac{r}{R}\right)^{1/m} = \left(\frac{\tilde{C}}{\theta R^{1/m}}\right) r^{1/m}$$

where $\tilde{C} = \dfrac{\alpha R\theta^{m-1}}{1 - \theta^{m-1}} + \omega(R)$. The conclusion follows with $C = \tilde{C}/\theta R^{1/m}$. □

Proof of Theorem 2.5.7 First of all, notice that under our assumptions, denoting by $x_{t_0,x_0,u_0}(.)$ the solution of (2.1) starting from x_0 at time t_0, associated with a control $u_0(.)$, by Gronwall's Lemma

$$\left|x_{t_0,x_0,u_0}(t)\right| \leq \left(|x_0| + (t - t_0)\,\|c\|_{\infty,[t_0,t]}\right) e^{(t-t_0)\|c\|_{\infty,[t_0,t]}} \qquad \forall t \geq t_0. \tag{2.71}$$

Hence, for every $R > 0$, $t > 0$, $t_0 \in [0, t]$, and all $x_0, x_1 \in B(0, R)$

$$\left|x_{t_0,x_1,u}(s) - x_{t_0,x_0,u}(s)\right| \leq |x_1 - x_0|\, e^{\int_{t_0}^{s} c_{M_{t_0,R}(t)}(\xi)\,d\xi} \qquad \forall s \in [t_0, t], \tag{2.72}$$

where we put $M_{t_0,R}(t) := \left(R + (t - t_0)\,\|c\|_{\infty,[t_0,t]}\right) e^{(t-t_0)\|c\|_{\infty,[t_0,t]}}$.

Let any $R > 0$ such that

$$A_R := \operatorname{dom} v \cap R\mathbb{B}$$

is nonempty. Then from our assumptions, v is bounded on A_R. We denote $x_{\tilde{x},u}(.) := x_{0,\tilde{x},u}(.)$ for any starting point $\tilde{x}$ and any control $u(.)$. Moreover, for any $\tilde{x} \in A$ we denote by $\mathscr{U}_{\tilde{x}}$ the set of all Lebesgue measurable functions $u : [0, 1] \to \mathbb{R}^m$ such that $u(t) \in U$ a.e. $t \geq 0$ and $x_{\tilde{x},u}(s) \in A$ for all $s \in [0, 1]$. By the dynamic programming principle it follows that for any distinct $x_1, x_0 \in A_R$ there exists $(x_{x_0,u_0}(.), u_0(.))$ solving (2.1), such that

$$v(x_0) + |x_1 - x_0| > \int_0^1 e^{-\lambda s} l(x_{x_0,u_0}(s), u_0(s))ds + e^{-\lambda} v(x_{x_0,u_0}(1)).$$

Thus, applying again the dynamic programming principle, it follows that for any $u_1 \in \mathscr{U}_{x_1}$

$$\begin{aligned}
&v(x_1) - v(x_0) \\
&\leq |x_1 - x_0| + \left| \int_0^1 e^{-\lambda s} \left(l(x_{x_1,u_1}(s), u_1(s)) - l(x_{x_0,u_0}(s), u_0(s)) \right) ds \right| \\
&\quad + e^{-\lambda} \left| v(x_{x_1,u_1}(1)) - v(x_{x_0,u_0}(1)) \right| \\
&\leq |x_1 - x_0| + \left| \int_0^1 e^{-\lambda s} \left(l(x_{x_1,u_1}(s), u_1(s)) - l(x_{x_1,u_0}(s), u_0(s)) \right) ds \right| \\
&\quad + \left| \int_0^1 e^{-\lambda s} \left[l(x_{x_1,u_0}(s), u_0(s)) - l(x_{x_0,u_0}(s), u_0(s)) \right] ds \right| \\
&\quad + e^{-\lambda} \left| v(x_{x_1,u_1}(1)) - v(x_{x_0,u_0}(1)) \right| .
\end{aligned} \tag{2.73}$$

By (2.71), there exists a constant $M \geq 0$ such that for all $\overline{x} \in A_R$ and all Lebesgue measurable $u : [0, 1] \to \mathbb{R}^m$ with $u(t) \in U$ a.e. t, the trajectories $x_{\overline{x},u}(.)$ take values in $B(0, M)$ on the time interval $[0, 1]$. Let $\tilde{C} > 0$ be a Lipschitz constant for $l(., .)$ on $B(0, M)$, with respect to the space variable. Then, by (2.72), there exists $c > 1$ such that for all $x_1, x_0 \in A_R$

$$\begin{aligned}
&\int_0^1 e^{-\lambda s} \left| l(x_{x_1,u_0}(s), u_0(s)) - l(x_{x_0,u_0}(s), u_0(s)) \right| ds \\
&\leq \tilde{C} \int_0^1 e^{-\lambda s} \left| x_{x_1,u_0}(s) - x_{x_0,u_0}(s) \right| ds \\
&\leq \tilde{C} \cdot c \, |x_1 - x_0| .
\end{aligned}$$

So, putting $C = \tilde{C} \cdot c + 1$, from (2.73) it follows that

$$\begin{aligned}
&v(x_1) - v(x_0) \\
&\leq C \, |x_1 - x_0| + \left| \int_0^1 e^{-\lambda s} \left(l(x_{x_1,u_1}(s), u_1(s)) - l(x_{x_1,u_0}(s), u_0(s)) \right) ds \right| \\
&\quad + e^{-\lambda} \left| v(x_{x_1,u_1}(1)) - v(x_{x_0,u_0}(1)) \right| .
\end{aligned} \tag{2.74}$$

Now we claim that there exist a constant $\beta = \beta(f, l) \geq 1$ and a control $u_1 \in \mathscr{U}_{x_1}$ such that

$$\left|\int_0^1 e^{-\lambda s}\left(l(x_{x_1,u_1}(s), u_1(s)) - l(x_{x_1,u_0}(s), u_0(s))\right) ds\right| \leq \beta\,|x_1 - x_0|\,, \qquad \left|x_{x_1,u_1}(1) - x_{x_0,u_0}(1)\right| \leq \beta\,|x_1 - x_0|\,. \tag{2.75}$$

Indeed, if $\max_{s\in[0,1]} d_A(x_{x_1,u_0}(s)) = 0$ then $u_0 \in \mathscr{U}_{x_1}$. So, (2.75) follows taking $u_1 = u_0$. Otherwise, suppose $\max_{s\in[0,1]} d_A(x_{x_1,u_0}(s)) > 0$ and consider the following control system in $\mathbb{R}^{n+1}$

$$\begin{cases} x'(s) = f(x(s), u(s)) & \text{a.e. } s \in [0, 1], \\ z'(s) = e^{-\lambda s} l(x(s), u(s)) & \text{a.e. } s \in [0, 1], \\ x(0) = \tilde{x},\ z(0) = 0, & \\ u(.) \text{ is Lebesgue measurable}, & \\ u(s) \in U & \text{a.e. } s \in [0, 1]. \end{cases} \tag{2.76}$$

Let us denote by $(X_{\tilde{x},u}, u)$ the trajectory-control pair that satisfies (2.76) where $X_{\tilde{x},u}(s) := (x_{\tilde{x},u}(s), \int_0^s e^{-\lambda\xi} l(x(\xi), u(\xi))d\xi)$. Set $\Omega := A \times \mathbb{R}$. By Theorem 2.3.6, there exists a constant $\beta \geq 1$ (depending only on f and l) and a control $u_1 \in \mathscr{U}_{x_1}$ such that

$$\left\|X_{x_1,u_1} - X_{x_1,u_0}\right\|_{\infty,[0,1]} \leq \beta\left(\max_{s\in[0,1]} d_\Omega(X_{x_1,u_0}(s))\right).$$

Since $d_\Omega(X_{x_1,u_0}(.)) = d_A(x_{x_1,u_0}(.))$ and $x_{x_0,u_0}(.) \in A$ we have

$$\begin{aligned} &\left\|X_{x_1,u_1} - X_{x_1,u_0}\right\|_{\infty,[0,1]} \\ &\leq \beta \max_{s\in[0,1]}\left\{\inf_{\gamma\in\Omega}\left|X_{x_1,u_0}(s) - \gamma\right|\right\} \\ &\leq \beta \max_{s\in[0,1]}\left\{\left|x_{x_1,u_0}(s) - x_{x_0,u_0}(s)\right|\right\} \\ &\leq \beta \cdot c\,|x_1 - x_0|\,. \end{aligned}$$

Furthermore

$$\left|\int_0^1 e^{-\lambda s}\left[l(x_{x_1,u_1}(s), u_1(s)) - l(x_{x_1,u_0}(s), u_0(s))\right] ds\right| \leq \left\|X_{x_1,u_1} - X_{x_1,u_0}\right\|_{\infty,[0,1]}\,,$$

$$\left|x_{x_1,u_1}(1) - x_{x_0,u_0}(1)\right| \leq \left\|X_{x_1,u_1} - X_{x_1,u_0}\right\|_{\infty,[0,1]} + c\,|x_1 - x_0|\,.$$

So, replacing β with $2\beta \cdot c$, (2.75) follows.

Combining the inequalities in (2.74) and (2.75) we obtain for all $0 \leq r \leq R/\beta$ and all $x_1, x_0 \in A_R$ with $|x_1 - x_0| \leq r$ that

$$v(x_1) - v(x_0) \leq (C + \beta)\, r + e^{-\lambda}\omega(\beta r),$$

where we defined

$$\omega(r) := \sup_{\substack{|x-\tilde{x}|\leq r \\ x,\tilde{x}\in A_R}} |v(x) - v(\tilde{x})| \quad \forall r \in [0, R].$$

By the symmetry of the previous inequality with respect to x_1 and x_0 we have that

$$|v(x_1) - v(x_0)| \leq (C + \beta)\, r + e^{-\lambda}\omega(\beta r). \tag{2.77}$$

Letting $\theta := e^{-\lambda}$ and $\alpha := C + \beta$, we deduce from (2.77) that for all $0 \leq r \leq R/\beta$

$$\omega(r) \leq \alpha r + \theta\, \omega(\beta r). \tag{2.78}$$

So, Lemma 2.5.9 yields the Lipschitz continuity of v on A_R for $\lambda > \log \beta$. In particular, taking into consideration that our assumptions yield to

$$V(t, x) = e^{-\lambda t} v(x) \quad \forall (t, x),$$

we get immediately the last statement. □

Remark 2.5.10 From (2.78) and Lemma 2.5.9 it follows that, given any $\lambda > 0$, $(V(t, .))_{t\in\mathbb{R}_0^+}$ are uniformly locally Hölder continuous on A of exponent $1/m$ for all $m \geq 1$ such that $m > (\log \beta)/\lambda$, where β is as in the above proof.

From Theorem 2.5.7, we infer the following corollary.

Corollary 2.5.11 *Assume A compact, $L(t, x, u) = e^{-\lambda t} l(t, x, u)$, and there exists $T > 0$ such that l is time independent for all $t \geq T$ and all the assumptions of Theorem 2.5.7 hold true. Then $(V(t, .))_{t\in\mathbb{R}_0^+}$ are uniformly Lipschitz continuous on A for all large $\lambda > 0$.*

Corollary 2.5.12 *Consider Assumption 2.5.1. Assume A compact and $L(t, x, u) = e^{-\lambda t} l(t, x, u)$ with the further conditions: $l(., x, u)$ is T-periodic, i.e. there exists $T > 0$ such that $l(t + T, x, u) = l(t, x, u)$ for all $t \in \mathbb{R}_0^+$, $x \in \mathbb{R}^n$ and $u \in \mathbb{R}^m$. Then $(V(t, .))_{t\in\mathbb{R}_0^+}$ are uniformly Lipschitz continuous on A for all large $\lambda > 0$.*

Proof Fix $t \in \mathbb{R}_0^+$. Then, by the dynamic programming principle, for any $x_1, x_0 \in A$ there exists u_0 feasible at x_0 such that

$$\begin{aligned} &V(t, x_1) - V(t, x_0) \\ &\leq |x_1 - x_0| + \left| \int_t^{t+T} e^{-\lambda s} \big(l(s, x_{x_1,u_1}(s), u_1(s)) - l(s, x_{x_0,u_0}(s), u_0(s)) \big)\, ds \right| \\ &\quad + \left| V(t + T, x_{x_1,u_1}(t + T)) - V(t + T, x_{x_0,u_0}(t + T)) \right| \end{aligned}$$

for any u_1 feasible at x_1. Now, the periodicity of l in the time variable implies that $V(s+T,x) = e^{-\lambda T} V(s,x)$. From the previous inequality it follows that

$$\begin{aligned}
&V(t,x_1) - V(t,x_0) \\
&\leq |x_1 - x_0| + \left| \int_t^{t+T} e^{-\lambda s} \left(l(s, x_{x_1,u_1}(s), u_1(s)) - l(s, x_{x_0,u_0}(s), u_0(s)) \right) ds \right| \\
&\quad + e^{-\lambda T} \left| V(t, x_{x_1,u_1}(t+T)) - V(t, x_{x_0,u_0}(t+T)) \right| .
\end{aligned}$$

Proceeding as in the proof of Theorem 2.5.7, by the neighbouring feasible trajectory result Theorem 2.3.6 there exist two constants $\beta \geq 1$ and $C \geq 0$ (depending only on f, l, and T) such that, for all $|x_1 - x_0| \leq r$, we have that

$$|V(t,x_1) - V(t,x_0)| \leq (C+\beta)r + e^{-\lambda T} \sup_{\substack{|h-\tilde{h}| \leq \beta r \\ h,\tilde{h} \in A}} \left| V(t,h) - V(t,\tilde{h}) \right| .$$

The conclusion follows applying Lemma 2.5.9 for $\lambda > (\log \beta)/T$. □

Chapter 3
First Order Necessary Conditions Under State Constraints

Abstract In this chapter, we present the normal maximum principle, complemented by sensitivity relations and an initial—time transversality condition. We prove the existence of accompanying co-state variables, associated measures, and a measurable selection that together satisfy the adjoint equations and optimality conditions. When the trajectory lies strictly within the admissible set, the standard maximum-principle relations hold. If instead the trajectory reaches the boundary of the constraint set, we introduce modified adjoint equations and boundary-adjusted optimality conditions.

Keywords Necessary conditions · Maximum principle · Transversality conditions · Sensitivity relations

3.1 Main Assumptions

Consider the infinite horizon optimal control problem $\mathscr{B}_\infty$

$$\text{minimize} \int_{t_0}^{+\infty} L(t, x(t), u(t))\, dt \tag{3.1}$$

over all the trajectory-control pairs subject to the state constrained control system

$$\begin{cases} x'(t) = f(t, x(t), u(t)) & \text{a.e. } t \in \mathbb{R}^+_{t_0}, \\ x(t_0) = x_0, & \\ u(t) \in U(t) & \text{a.e. } t \in \mathbb{R}^+_{t_0}, \\ x(t) \in A & \forall t \in \mathbb{R}^+_{t_0}, \end{cases} \tag{3.2}$$

where $f : \mathbb{R}^+_0 \times \mathbb{R}^n \times \mathbb{R}^m \to \mathbb{R}^n$ and $L : \mathbb{R}^+_0 \times \mathbb{R}^n \times \mathbb{R}^m \to \mathbb{R}$ are given, A is a nonempty closed subset of $\mathbb{R}^n$, $U : \mathbb{R}^+_0 \rightsquigarrow \mathbb{R}^m$ is a Lebesgue measurable set valued map with closed nonempty images and $(t_0, x_0) \in \mathbb{R}^+_0 \times A$ is the initial datum. Every trajectory-control pair $(x(.), u(.))$ that satisfies the state constrained control system (3.2) is called feasible. We refer to such $x(.)$ as a feasible trajectory. The infimum

V. Basco, *Nonsmooth Constrained Optimal Control on Infinite Horizon*, SpringerBriefs in Mathematics, https://doi.org/10.1007/978-3-032-08938-0_3

of the cost functional in (3.1) over all feasible trajectory-control pairs, with the initial datum (t_0, x_0), is denoted by $V(t_0, x_0)$ (if no feasible trajectory-control pair exists at (t_0, x_0), or if the integral in (3.1) is not defined for every feasible pair, we set $V(t_0, x_0) = +\infty$). The function $V : \mathbb{R}_0^+ \times A \to [-\infty, +\infty]$ is called the value function of problem $\mathscr{B}_\infty$. We define the Hamiltonian function on $\mathbb{R}_0^+ \times \mathbb{R}^n \times \mathbb{R}^n$ by

$$H(t, x, p) = \sup\{\langle p, f(t, x, u)\rangle - L(t, x, u) \mid u \in U(t)\}.$$

Assumption 3.1.1 We consider the following assumptions on f and L:

(I) there exist two locally essentially bounded functions $b, \theta : \mathbb{R}_0^+ \to \mathbb{R}_0^+$ and a nondecreasing function $\Psi : \mathbb{R}_0^+ \to \mathbb{R}_0^+$ such that for a.e. $t \in \mathbb{R}_0^+$ and for all $x \in \mathbb{R}^n$, $u \in U(t)$

$$|f(t, x, u)| \leq b(t)\,(1 + |x|)\,,$$
$$|L(t, x, u)| \leq \theta(t)\Psi(|x|);$$

(II) for any $R > 0$ there exist c_R, $\alpha_R \in L^1_{\mathrm{loc}}(\mathbb{R}_0^+; \mathbb{R}_0^+)$ such that for a.e. $t \in \mathbb{R}_0^+$ and for all x, $y \in B(0, R)$, $u \in U(t)$,

$$|f(t, x, u) - f(t, y, u)| \leq c_R(t)|x - y|,$$
$$|L(t, x, u) - L(t, y, u)| \leq \alpha_R(t)\,|x - y|\,;$$

(III) for all $x \in \mathbb{R}^n$ the mappings $f(., x, .)$, $L(., x, .)$ are Lebesgue-Borel measurable;

(IV) for a.e. $t \in \mathbb{R}_0^+$, and for all $x \in \mathbb{R}^n$ the set

$$\{(f(t, x, u), L(t, x, u)) \mid u \in U(t)\}$$

is closed;

(V) for all $(t_0, x_0) \in \mathbb{R}_0^+ \times A$ the limit $\lim_{T\to+\infty} \int_{t_0}^T L(t, x(t), u(t))\,dt$ exists for all trajectory-control pairs (x, u) satisfying (3.2) with initial datum (t_0, x_0);

(VI) $V(t_0, x_0) \neq -\infty$ for all $(t_0, x_0) \in \mathbb{R}_0^+ \times A$.

Remark 3.1.2 A sufficient condition to guarantee that the last two conditions in Assumption 3.1.1 are satisfied is to assume that L takes nonnegative values. Alternatively, we may assume that for any initial datum (t_0, x_0) there exists a function $\phi_{t_0,x_0} \in L^1(\mathbb{R}_0^+; \mathbb{R}_0^+)$ such that $L(t, x(t), u(t)) \geq \phi_{t_0,x_0}(t)$ a.e. $t \in \mathbb{R}_{t_0}^+$ for all trajectory-control pairs (x, u) satisfying (3.2).

3.1.1 An Illustrative Example

In mathematical economics, it is common to explore trajectories that satisfy both the unconstrained Pontryagin maximum principle and state constraints when seeking optimal solutions. Such an approach, however, is incorrect as there are cases (see example below) where no optimal trajectory exists in this class. There is, therefore, the need of a constrained maximum principle for infinite horizon problems with sufficiently general structure. Indeed, in the next example we will show the fallacy of applying the unconstrained Pontryagin maximum principle to $\mathscr{B}_\infty$ in order to obtain candidates for optimality that satisfy some given state constraints.

Consider the following infinite horizon optimal control problem:

$$\text{maximize } J(u) = \int_0^{+\infty} e^{-\lambda t}(x(t) + u(t))\, dt$$

over all measurable controls $u(t) \in [-1, 1]$ a.e. $t \in \mathbb{R}_0^+$, where $x(.)$ is the associated locally absolutely continuous solution of

$$\begin{cases} x'(t) = -au(t) & \text{a.e. } t \in \mathbb{R}_0^+, \\ x(0) = 1, \\ x(t) \in]-\infty, 1] & \forall t \in \mathbb{R}_0^+, \end{cases}$$

with $a > \lambda > 0$. Applying the Pontryagin maximum principle for unconstrained problems, it follows that any optimal trajectory-control pair satisfies one of the following three relations:

(i) $x^-(t) = 1 + at$ associated with $u^-(t) \equiv -1$;
(ii) $x^+(t) = 1 - at$ associated with $u^+(t) \equiv +1$;
(iii) $x^\pm(t) = (1 - at)\chi_{[0,\bar{t}]}(t) + (1 - a\bar{t} + a(t - \bar{t}))\chi_{]\bar{t},+\infty[}(t)$ associated with $u^\pm(t) = \chi_{[0,\bar{t}]}(t) - \chi_{]\bar{t},+\infty[}(t)$, for some $\bar{t} > 0$.

Excluding now the trajectories x^- and $x^\pm$, since they are not feasible, this analysis leads to the conclusion that x^+ is the only candidate for optimality. But one can easily see that the feasible trajectory $\bar{x}(t) \equiv 1$, associated with the control $\bar{u}(t) \equiv 0$, verifies

$$J(\bar{u}) > J(u^+).$$

3.2 Finite Horizon Reductions

A fundamental technique for analyzing infinite horizon problems consists in comparing them with suitable finite horizon approximations. This approach allows transferring known results from classical theory to the infinite case and developing numerical approximation schemes. Here, we show how properties of the original problem are reflected in its finite horizon reductions.

Notice that the Assumption 3.1.1 guarantee the existence and uniqueness of the solution to the differential equation in (3.2) for every initial datum x_0 and every control.

Define the extended value function $V : \mathbb{R}_0^+ \times A \to]-\infty, +\infty]$ of problem $\mathscr{B}_\infty$ by

$$V(t_0, x_0) := \inf \left\{ \int_{t_0}^{+\infty} L(t, x(t), u(t))\, dt \mid (x(.), u(.)) \text{ sol. of (3.2)}, x(t_0) = x_0 \right\}.$$

We say that a pair $(\bar{x}(.), \bar{u}(.))$ is *optimal* for $\mathscr{B}_\infty$ at $(t_0, x_0) \in \text{dom}\, V$ if

$$\int_{t_0}^{+\infty} L(t, \bar{x}(t), \bar{u}(t))\, dt \le \int_{t_0}^{+\infty} L(t, x(t), u(t))\, dt$$

for any feasible trajectory-control pair $(x(.), u(.))$ starting from x_0 at time t_0.

Lemma 3.2.1 *Let $T \ge 0$ and consider Assumption 3.1.1. Let $\mathscr{B}_T$ be the Bolza problem*

$$\text{minimize} \left\{ V(T, x(T)) + \int_{t_0}^{T} L(t, x(t), u(t))\, dt \right\}$$

over all the trajectory-control pairs satisfying the state constrained equation

$$\begin{cases} x'(t) = f(t, x(t), u(t)) & a.e.\ t \in [t_0, T], \\ x(t_0) = x_0, & \\ u(t) \in U(t) & a.e.\ t \in [t_0, T], \\ x(t) \in A & t \in [t_0, T]. \end{cases}$$

Denote by $V_{\mathscr{B}_T} : [0, T] \times A \to]-\infty, +\infty]$ the value function of the above problem. Then

$$V_{\mathscr{B}_T}(., .) = V(., .) \quad on\ [0, T] \times A. \tag{3.3}$$

Furthermore, if $(\bar{x}(.), \bar{u}(.))$ is optimal at $(t_0, x_0) \in [0, T] \times A$ for $\mathscr{B}_\infty$, then the restriction of $(\bar{x}, \bar{u})$ to the time interval $[t_0, T]$ is optimal for the Bolza problem $\mathscr{B}_T$ too.

Proof Let $(t_0, x_0) \in [0, T] \times A$ and $\varepsilon > 0$. If $V(t_0, x_0) = +\infty$, then $V(t_0, x_0) \ge V_{\mathscr{B}_T}(t_0, x_0)$. Otherwise, there exists a feasible trajectory-control pair $(x_\varepsilon, u_\varepsilon)$ for problem $\mathscr{B}_\infty$ at (t_0, x_0) such that

$$\begin{aligned} V(t_0, x_0) &\ge \int_{t_0}^{T} L(s, x_\varepsilon(s), u_\varepsilon(s))ds + \int_{T}^{+\infty} L(s, x_\varepsilon(s), u_\varepsilon(s))ds - \varepsilon \\ &\ge \int_{t_0}^{T} L(s, x_\varepsilon(s), u_\varepsilon(s))ds + V(T, x_\varepsilon(T)) - \varepsilon \\ &\ge V_{\mathscr{B}_T}(t_0, x_0) - \varepsilon. \end{aligned} \tag{3.4}$$

Since ε is arbitrary, we obtain $V(t_0, x_0) \ge V_{\mathscr{B}_T}(t_0, x_0)$.

On the other hand, if $V_{\mathscr{B}_T}(t_0, x_0) = +\infty$, then $V_{\mathscr{B}_T}(t_0, x_0) \geq V(t_0, x_0)$. Otherwise, there exists a feasible trajectory-control pair $(\tilde{x}_\varepsilon, \tilde{u}_\varepsilon)$ for problem $\mathscr{B}_T$ at (t_0, x_0) such that

$$V_{\mathscr{B}_T}(t_0, x_0) \geq \int_{t_0}^{T} L(s, \tilde{x}_\varepsilon(s), \tilde{u}_\varepsilon(s))ds + V(T, \tilde{x}_\varepsilon(T)) - \varepsilon.$$

By (2.71) and our assumptions on L, $\int_{t_0}^{T} L(s, \tilde{x}_\varepsilon(s), \tilde{u}_\varepsilon(s))ds < +\infty$. Hence $(T, \tilde{x}_\varepsilon(T)) \in \text{dom}\, V$. So, there exists a feasible trajectory-control pair $(\hat{x}_\varepsilon, \hat{u}_\varepsilon)$ for problem $\mathscr{B}_\infty$ at $(T, \tilde{x}_\varepsilon(T))$ such that

$$\begin{aligned} V_{\mathscr{B}_T}(t_0, x_0) &\geq \int_{t_0}^{T} L(s, \tilde{x}_\varepsilon(s), \tilde{u}_\varepsilon(s))ds + \int_{T}^{+\infty} L(s, \hat{x}_\varepsilon(s), \hat{u}_\varepsilon(s))ds - 2\varepsilon \\ &= \int_{t_0}^{+\infty} L(s, x(s), u(s))ds - 2\varepsilon, \end{aligned} \tag{3.5}$$

where $x(.)$ is the trajectory starting from x_0 at time t_0 satisfying the ordinary differential equation in (3.2) with the control u given by

$$u(s) := \begin{cases} \tilde{u}_\varepsilon(s) & s \in [t_0, T], \\ \hat{u}_\varepsilon(s) & s \in]T, +\infty[. \end{cases}$$

Since $u(.) \in U(.)$ and $x(\mathbb{R}^+_{t_0}) \subset A$, (x, u) is feasible for problem $\mathscr{B}_\infty$ at (t_0, x_0). Then, by (3.5), $V_{\mathscr{B}_T}(t_0, x_0) \geq V(t_0, x_0) - 2\varepsilon$ and, since ε is arbitrary, $V_{\mathscr{B}_T}(t_0, x_0) \geq V(t_0, x_0)$.

The last part of the conclusion follows from (3.4), by setting $\varepsilon = 0$, $(x_\varepsilon, u_\varepsilon) = (\bar{x}, \bar{u})$, and using that $V_{\mathscr{B}_T}(t_0, x_0) = V(t_0, x_0)$. □

3.2.1 *Set Constraints Reduction*

This subsection presents a way for reducing problems formulated with set constraints to equivalent functional formulations that are more tractable from both analytical and computational perspectives, while preserving optimality.

Consider the following optimal control problem

$$\text{minimize } \phi(x(T))$$

over all $x \in W^{1,1}([S, T]; \mathbb{R}^n)$ such that

$$\begin{cases} x'(s) \in \mathscr{F}(s, x(s)) & \text{a.e. } s \in [S, T], \\ x(s) \in A & \forall t \in [S, T], \\ x(S) = x_0, \end{cases} \tag{3.6}$$

where $S \in [\tau, T], 0 \leq \tau < T, x_0 \in \mathbb{R}^n, \phi : \mathbb{R}^n \to \mathbb{R}, \mathscr{F} : [\tau, T] \times \mathbb{R}^n \rightsquigarrow \mathbb{R}^n$, and a closed set $A \subset \mathbb{R}^n$ are given. We define the value function by

$$v(S, x_0) = \inf\{\phi(x(T)) \mid x \text{ solution of (3.6) with initial datum } (S, x_0)\}.$$

In this section we prove the following result.

Proposition 3.2.2 *Let $\bar{x}(.)$ be a minimizer at (S, x_0). Assume that:*

- *ϕ is Lipschitz continuous on a neighborhood of $\bar{x}(T)$;*
- *$\mathscr{F}(t, x) \neq \emptyset$ for all (t, x), $\mathscr{F}(., .)$ is Lebesgue-Borel measurable,* graph $\mathscr{F}(t, .)$ *is closed for each $t \in [\tau, T]$, and there exist $\varepsilon > 0$ and $k \in L^1([\tau, T]; [0 + \infty[)$ such that, for a.e. $t \in [\tau, T]$,*

$$\mathscr{F}(t, \tilde{x}) \subset \mathscr{F}(t, x) + k(t)\,|\tilde{x} - x|\,\mathbb{B},$$

for all $x, \tilde{x} \in \bar{x}(t) + \varepsilon\mathbb{B}$.

Moreover, assume the following controllability conditions:

- *there exists $q \in L^1([\tau; T]; \mathbb{R}_0^+)$ such that $\mathscr{F}(t, x) \subset q(t)\,\mathbb{B}$ for all $x \in$ bdr A, and a.e. $t \in [\tau, T]$;*
- *there exist $\eta > 0$, $r > 0$, and $M \geq 0$ such that: for a.e. $t \in [\tau, T]$, all $y \in$ bdr $A + \eta\mathbb{B}$, and all*

$$v \in \mathscr{F}(t, y) \cap \{p \in \mathbb{R}^n \mid \exists n \in N_A(x; \eta),\ \langle p, n\rangle \geq 0\},$$

there exists $w \in \mathscr{F}(t, y) \cap B(v, M)$ such that

$$w,\ w - p \in \{p \in \mathbb{R}^n \mid \langle p, n\rangle \leq -r,\ \forall n \in N_A(x; \eta)\},$$

where $N_A(x; \eta)$ is defined in (2.38).

Then, for any sufficiently small $\varepsilon > 0$ there exist $\delta > 0$, $K > 0$, and functions $\zeta^\varepsilon : [S, T] \times \mathbb{R}^n \to \mathbb{R}$, $\tilde{\zeta}^\varepsilon : [S, T] \times \mathbb{R}^n \times \mathbb{R} \to \mathbb{R}$, with $\zeta^\varepsilon(., x)$, $\tilde{\zeta}^\varepsilon(., y, r)$ Lebesgue measurable for all $x, y \in \mathbb{R}^n$, $r \in \mathbb{R}$ and $\zeta^\varepsilon(t, .)$, $\tilde{\zeta}^\varepsilon(t, ., .)$ K-Lipschitz continuous, such that the trajectory $\bar{X}(.) := (\bar{x}(.), \bar{a}(.) \equiv 0, \bar{b}(.) \equiv 0, \bar{c}(.) \equiv 0)$ is a δ-minimizer for

$$\textit{minimize } \phi(x(T)) - v(S, x(S)) + \int_S^T \zeta^\varepsilon(t, a(t))dt + \int_S^T \tilde{\zeta}^\varepsilon(t, b(t), c(t))dt \tag{3.7}$$

over all $(x, a, b) \in \left(W^{1,1}([S,T];\mathbb{R}^n)\right)^3$ *and* $c \in W^{1,1}([S,T];\mathbb{R})$ *such that*

$$\begin{cases} x'(t) \in a(t) + (1 + c(t))(b(t) + \mathscr{F}(t, x(t))) & a.e.\ t \in [S,T], \\ (a'(t), b'(t), c'(t)) \in (\varepsilon\mathbb{B})^3 & a.e.\ t \in [S,T], \\ x(t) \in A \cap (\bar{x}(t) + (\varepsilon/4)\mathbb{B}) & \forall t \in [S,T], \\ (a(S), b(S), c(S)) = 0. \end{cases} \tag{3.8}$$

We recall that, with δ-minimizer we mean $\left\|\bar{X} - X\right\|_{\infty,[S,T]} \le \delta$ over all $X = (x(.), a(.), b(.), c(.))$ trajectory solving (3.8).

Let us show some preliminary results.

Lemma 3.2.3 *The value function* $v(.,.)$ *is locally Lipschitz continuous on* $[\tau, T] \times A$*. Moreover, there exists a subset* $\mathcal{C} \subset [\tau, T]$*, with* $\mu_{\mathscr{L}}(\mathcal{C}) = 0$*, such that* $D_{\uparrow}v(t, x)(1, f) \ge 0$ *for all* $t \in [\tau, T]\backslash\mathcal{C}$*, all* $x \in$ *int* A*, and all* $f \in \mathscr{F}(t, x)$*.*

Proof From the proof of Theorem 2.3.6 and our assumptions, it follows the neighboring feasible trajectory estimates as in the conclusion of Theorem 2.3.6 on the compact interval $[\tau, T]$. So, using Lemma 2.5.9, the locally Lipschitz continuity of $\phi(.)$, and arguing as in the proof of Theorem 2.5.4, we can conclude[1] that $v(.,.)$ is locally Lipschitz continuous on $[\tau, T] \times A$. Moreover, from Lemma 2.2.5, there exists a set $\mathcal{C} \subset [\tau, T]$ with null measure such that for all $t \in]\tau, T[\backslash\mathcal{C}$, all $x \in \text{int}\, A$, and all $f \in \mathscr{F}(t, x)$ there exists a solution $x'(t) = \mathscr{F}(t, x(t))$ for a.e. t satisfying $x'(t) = f$. Since $x(t) \in \text{int}\, A$ and from the dynamical programming principle, we have $x(t + h) \in \text{int}\, A$ and $v(t + h, x(t + h)) \ge v(t, x(t))$ for all small $h > 0$. Then, using that v is locally Lipschitz, dividing by h and passing to the lower limit as $h \to 0+$, we get also the second statement. □

Let $\bar{x}(.)$ be a minimizer. For any $\varepsilon > 0$ we define $\zeta^{\varepsilon} : [S, T] \times \mathbb{R}^n \to \mathbb{R}$ and $\tilde{\zeta}^{\varepsilon} : [S, T] \times \mathbb{R}^n \times \mathbb{R} \to \mathbb{R}$ by

$$\begin{aligned} \zeta^{\varepsilon}(t, a) &:= \sup\left\{\langle a, \xi\rangle \mid \xi \in -\partial_x v(t, x), x \in (\bar{x}(t) + \varepsilon\mathbb{B}) \cap \text{int}\, A\right\}, \\ \tilde{\zeta}^{\varepsilon}(t, b, c) &:= \sup\{\langle (\xi, \eta), (c, -(1 + c)b)\rangle \mid (\xi, \eta) \in \partial v\left(\tilde{t}, \tilde{x}\right), \\ &\qquad (\tilde{t}, \tilde{x}) \in ((t, \bar{x}(t)) + \varepsilon\mathbb{B}) \cap ([S, T] \times \text{int}\, A)\}. \end{aligned}$$

Notice that,

$$\begin{aligned} \zeta^{\varepsilon}(t, a) &= \sup\left\{\langle \xi, a\rangle \mid \xi \in D\right\}, \\ \tilde{\zeta}^{\varepsilon}(t, b, c) &= \sup\left\{\langle (\xi, \eta), (c, -(1 + c)b)\rangle \mid (\xi, \eta) \in \tilde{D}\right\}, \end{aligned}$$

where

[1] We notice that in Theorem 2.5.4 a discounted Lagrangian over an infinite horizon is assumed; an analogous proof also applies to Mayer problems with minor adjustments.

$$D := \{\xi \mid \xi \in -\mathrm{co}\, \partial_x v(t,x),\, x \in (\bar{x}(t) + \varepsilon\mathbb{B}) \cap \mathrm{int}\, A\},$$
$$\tilde{D} := \{(\xi,\eta) \mid (\xi,\eta) \in \mathrm{co}\, \partial v\left(\tilde{t},\tilde{x}\right),\, (\tilde{t},\tilde{x}) \in ((t,\bar{x}(t)) + \varepsilon\mathbb{B}) \cap ([S,T] \times \mathrm{int}\, A)\}.$$

We need the following result.

Lemma 3.2.4 *Let $(\Theta, \mathfrak{A}, \mu)$ be a complete σ-finite measure space. Consider two complete separable metric spaces X, Y, a measurable set-valued map $\Phi : \Theta \rightsquigarrow X$ with closed nonempty images, and a Carathéodory function*[2] *$\phi : \Theta \times X \to \mathbb{R}$. Then the marginal function $\alpha : \Theta \to [-\infty, +\infty[$ defined by*

$$\forall \theta \in \Theta, \quad \alpha(\theta) = \inf_{x \in \Phi(\theta)} \phi(\theta, x)$$

is measurable. Furthermore, the marginal map Ξ defined by

$$\forall \theta \in \Theta, \quad \Xi(\theta) = \{x \in \Phi(\theta) \mid \phi(\theta,x) = \inf_{y \in \Phi(\theta)} \phi(\theta, y)\}$$

is also measurable.

Proof See Sect. A.3.2. □

From Lemma 3.2.4, the functions $\zeta^{\varepsilon}(.,x)$ and $\tilde{\zeta}^{\varepsilon}(.,y,r)$ are Lebesgue measurable for all $x, y \in \mathbb{R}^n$, $r \in \mathbb{R}$. Furthermore, from Lemma 3.2.3 and the properties of limiting subdifferentials and polar functions, $\zeta^{\varepsilon}(t,.)$ and $\tilde{\zeta}^{\varepsilon}(t,.,.)$ are Lipschitz continuous with a suitable constant $K > 0$ not depending on time t.

Lemma 3.2.5 *Let $\varepsilon > 0$ and $S \le t_0 < t_1 \le T$. Consider $b : [S,T] \to \mathbb{R}^n$, $c : [S,T] \to \mathbb{R}$ bounded measurable functions and let $y : [t_0,t_1] \to \mathrm{int}\, A$ be a piecewise absolutely continuous function, continuous from the right, with $s \in]t_0,t_1[$ unique discontinuity point. Assume that $y|_{[t_0,s[}$ and $y|_{[s,t_1]}$ satisfy*

$$y'(t) \in (1 + c(t))(b(t) + \mathcal{F}(t, y(t))) \text{ a.e. } t \quad \& \quad \|\bar{x} - y\|_{\infty,[t_0,t_1]} < \varepsilon.$$

If $\left[y\left(s^-\right), y(s)\right] \subset \mathrm{int}\, A$ then

$$v\left(t_1, y\left(t_1\right)\right) - v\left(t_0, y\left(t_0\right)\right) \ge -\zeta^{\varepsilon}\left(s, y(s) - y\left(s^-\right)\right) - \int_{t_1}^{t_2} \tilde{\zeta}^{\varepsilon}(t, b(t), c(t))dt$$

where $y(s^-) := \lim_{t \to s-} y(t)$.

Proof We can write

$$\begin{aligned} &v\left(t_1, y\left(t_1\right)\right) - v\left(t_0, y\left(t_0\right)\right) \\ &= v(t_1, y\left(t_1\right)) - v(s, y(s)) + v(s, y(s)) - v\left(s, y\left(s^-\right)\right) \\ &\quad + v\left(s, y\left(s^-\right)\right) - v\left(t_0, y\left(t_0\right)\right). \end{aligned}$$

[2] A function $\phi : \Theta \times X \to \mathbb{R}$ is called a *Carathéodory function* if $\phi(\theta,.)$ is continuous for all θ and $\phi(.,x)$ is measurable (wrt the measure μ) for every x.

Now, consider the subset $\mathcal{C} \subset [S, T]$ of Lemma 3.2.3 and a measurable selection $w(t) \in \mathscr{F}(t, y(t))$ such that

$$y'(t) = (1 + c(t))(b(t) + w(t)) \text{ a.e. } t. \tag{3.9}$$

Take $t \in [S, T] \backslash \mathcal{C}$ at which the Lipschitz continuous function $v(., y(.))$ is differentiable, $w(t) \in \mathscr{F}(t, y(t))$, and (3.9) hold. We have

$$\frac{d}{dt}v(t, y(t)) = \lim_{h \to 0+} \frac{v(t + h, y(t) + h(1 + c(t))(w(t) + b(t))}{h} - v(t, y(t))).$$

By the Mean Value Theorem, for all sufficiently small $h > 0$ and a suitable $\sigma_h \in [0, 1]$, we can find

$$p_h \in \text{co}\, \partial v\, (t + h + h\sigma_h c(t), y(t) + h(1 + c(t))(v(t) + b(t)) - h\sigma_h(1 + c(t))b(t))$$

satisfying

$$\begin{aligned} \langle hp_h, (c(t), -(1 + c(t))b(t)) \rangle &= \Delta v(h) \\ &:= v(t + h(1 + c(t)), y(t) + h(1 + c(t))w(t)) \\ &\quad - v(t + h, y(t) + h(1 + c(t))(b(t) + w(t))). \end{aligned}$$

From Lemma 3.2.3, $D_{\uparrow}v(t, y(t))(1, w(t)) \geq 0$. Since $D_{\uparrow}v(t, y(t))(.)$ is positively homogeneous, it follows that $D_{\uparrow}v(t, y(t))(1 + c(t), (1 + c(t))w(t)) \geq 0$. Hence

$$\begin{aligned} 0 &\leq \liminf_{h \to 0+} \frac{v(t + h(1 + c(t)), y(t) + h(1 + c(t))w(t)) - v(t, y(t))}{h} \\ &\leq \limsup_{h \to 0+} \frac{v(t + h, y(t) + h(1 + c(t))(w(t) + b(t))) - v(t, y(t))}{h} + \limsup_{h \to 0+} \frac{\Delta v(h)}{h} \\ &\leq \frac{d}{dt}v(t, y(t)) + \limsup_{h \to 0+} \langle\, p_h, (c(t), -(1 + c(t))b(t)) \rangle. \end{aligned}$$

We conclude that $\frac{d}{dt}v(t, y(t)) + \tilde{\zeta}^{\varepsilon}(t, b(t), c(t)) \geq 0$, by definition of $\tilde{\zeta}^{\varepsilon}$. It follows that

$$v\,(t_1, y\,(t_1)) - v(s, y(s)) \geq -\int_s^{t_1} \tilde{\zeta}^{\varepsilon}(t, b(t), c(t))dt$$

and

$$v\left(s, y\left(s^-\right)\right) - v\,(t_0, y\,(t_0)) \geq -\int_{t_0}^{s} \tilde{\zeta}^{\varepsilon}(t, b(t), c(t))dt.$$

Recalling that $\left[y(s), y\left(s^{-}\right)\right] \subset (\bar{x}(s)+\varepsilon\mathbb{B}) \cap \operatorname{int} A$, we deduce from the Mean Value Theorem that

$$v(s, y(s)) - v\left(s, y\left(s^{-}\right)\right) = \left\langle \xi, \left(y(s) - y\left(s^{-}\right)\right)\right\rangle$$

for some $\xi \in \operatorname{co} \partial_x v(s, x)$, where x is a point in $\left[y\left(s^{-}\right), y(s)\right]$. This implies that

$$v(s, y(s)) - v\left(s, y\left(s^{-}\right) \geq -\zeta^{\varepsilon}\left(s, y(s) - y\left(s^{-}\right)\right).$$

Combining these relations yields the desired inequality. □

Lemma 3.2.6 *Let $q \in L^{\infty}([S, T]; \mathbb{R}^n)$. Then there exists a subset $\mathscr{S} \subset [S, T]$, with full measure, and a sequence $(N_j)_j \subset \mathbb{R}_0^+$ such that*

$$\lim_{j\to+\infty} \sup_{t\in[S,T]} \left| \int_S^t (\tilde{q}_{N_{j,\tau}}(s) - q(s))ds \right| = 0 \quad \forall \tau \in \mathscr{S}, \tag{3.10}$$

$$\lim_{j\to+\infty} \sup_{t\in[S,T]} \left| \int_S^t dq_{N_{j,\tau}}(s) - \int_0^t q(s)ds \right| = 0 \quad \forall \tau \in \mathscr{S}, \tag{3.11}$$

where we defined:

- $\tilde{q}_{N,\tau}(s) := q(\tau + \psi_N(s-\tau))$ *with* $\psi_N(s) := \sum_{j=-\infty}^{+\infty} \frac{j}{N} \chi_{[j/N,(j+1)/N[}(s)$;
- $q_{N,\tau} := \frac{1}{N} \sum_{j=-\infty}^{+\infty} q\left(\tau + \frac{j}{n}\right) \delta_{\{j/N+\tau\}}$;
- $\delta_{\{t\}}$ *the Dirac measure with mass at t.*

Proof See Sect. A.3.2. □

Let $\varepsilon > 0$ and $(x(.), a(.), b(.), c(.))$ be an arbitrary feasible trajectory for (3.7). Consider a closed bounded set $\mathscr{G} \subset \mathbb{R}^n$. From Lemma 1.1.2:*(iv)* and a compactness argument, there exists $\varepsilon > 0$ such that

$$\operatorname{int} T_A^C(z) \cap \left(\liminf_{(s,y) \underset{[S,T]\times A}{\longrightarrow} (t,z)} (a(s) + \operatorname{co}(1 + c(s))(b(s) + \mathscr{F}(s, y))) \right) \neq \emptyset$$

for all $(t, z) \in [S, T] \times (\mathscr{G} \cap \operatorname{bdr} A)$. From the results on distance estimates in Sect. 2.3.1, we can find a sequence of absolutely continuous functions $x_i : [S, T] \to \mathbb{R}^n$ such that, for each i, $x_i(S) = x(S)$ and

$$\begin{cases} x_i'(t) \in a(t) + (1 + c(t))(b(t) + \mathscr{F}(t, x_i(t))) & \text{a.e. } t \in [S, T], \\ x_i(t) \in \operatorname{int} A & \forall t \in]S, T], \end{cases}$$

$$\lim_{i\to+\infty} \|x_i - x\|_{\infty,[S,T]} = 0. \tag{3.12}$$

Since $\|\bar{x} - x\|_{\infty,[S,T]} \leq \varepsilon/4$, we can assume that

$$\|\bar{x} - x_i\|_{\infty,[S,T]} \leq \varepsilon/3 \quad \forall i \in \mathbb{N}.$$

Take a sequence $\delta_i \to 0+$. Now fix $i \in \mathbb{N}$, and keep $\rho > 0$ such that

$$x_i(t) + \rho\mathbb{B} \subset \operatorname{int} A \quad \forall t \in [S + \delta_i, T]. \tag{3.13}$$

Now, using the same notation as in the statement of Lemma 3.2.6, for any given $\tau \in \mathbb{R}$ and $N \in \mathbb{N}$, let us consider the discrete measures on Borel subsets of $[S + \delta_i, T]$ associated with the bounded measurable functions $s \mapsto a(s)$ and $s \mapsto \vartheta^{\varepsilon}(s) := \zeta^{\varepsilon}(s, a(s))$:

$$a_{N,\tau} = \frac{1}{N} \sum_{j=-\infty}^{+\infty} a\left(\tau + \frac{j}{N}\right) \delta_{\{\tau + \frac{j}{N}\}}$$

and

$$\begin{aligned} \vartheta^{\varepsilon}_{N,\tau} &= \frac{1}{N} \sum_{j=-\infty}^{+\infty} \vartheta^{\varepsilon}\left(\tau + \frac{j}{N}\right) \delta_{\{\tau + \frac{j}{N}\}} \\ &= \frac{1}{N} \sum_{j=-\infty}^{+\infty} \zeta^{\varepsilon}\left(\tau + \frac{j}{N}, a\left(\tau + \frac{j}{N}\right)\right) \delta_{\{\tau + \frac{j}{N}\}}. \end{aligned} \tag{3.14}$$

Applying Lemma 3.2.6 to the $\mathbb{R}^{n+1}$ valued function $s \mapsto (a(s), \vartheta^{\varepsilon}(s))$, there exists $\tau \in [S + \delta_i, T]$ such that $S + \delta_i - \tau$ and $T - \tau$ are irrational, and a sequence of integers $(N_j)_j \subset \mathbb{R}_0^+$ such that

$$\lim_{j \to +\infty} \int_{[S+\delta_i,T]} d\vartheta^{\varepsilon}_{N_j,\tau}(s) = \int_{[S+\delta_i,T]} \zeta^{\varepsilon}(s, a(s))ds$$

and

$$\lim_{j \to +\infty} \sup_{t \in [S+\delta_i,T]} \left| \int_{[S+\delta_i,t]} da_{N_j,\tau}(s) - \int_{[S+\delta_i,t]} a(s)ds \right| = 0. \tag{3.15}$$

Note next that the trajectory $x_i(.)$ satisfies

$$x_i(t) := \int_{[S+\delta_i,t]} a(s)ds + y_i(t) \quad \forall t \in [S + \delta_i, T],$$

where $y_i(.)$ is a solution to the differential inclusion

$$\begin{cases} y_i'(t) \in (1 + c(t)) \left(b(t) + \mathscr{F}\left(t, y_i(t) + \int_{[S+\delta_i,t]} a(s)ds\right)\right) \text{ a.e. } t, \\ y_i\left(S + \delta_i\right) = x_i\left(S + \delta_i\right). \end{cases}$$

We deduce from the Generalized Filippov Existence Theorem that, there exists a sequence of solution $(y_j)_j$ to the differential inclusion

$$\begin{cases} y_j'(t) \in (1+c(t))\left(b(t)+\mathscr{F}\left(t, y_j(t)+\int_{[S+\delta_i,t]} da_{N_j,\tau}(s)\right)\right) & \text{a.e. } t \\ y_j(S+\delta_i) = x_i(S+\delta_i) \end{cases}$$

such that

$$\lim_{j\to+\infty} \|y_j - y_i\|_{\infty,[S+\delta_i,T]} = 0. \tag{3.16}$$

From of (3.15) and (3.16) we deduce

$$\lim_{j\to+\infty} \sup_{t\in[S+\delta_i,T]} \left| \int_{[S+\delta_i,t]} da_{N_j,\tau}(s) + y_j(t) - \int_{[S+\delta_i,t]} a(s)ds + y_i(t) \right| = 0. \tag{3.17}$$

Now, letting for every $j \in \mathbb{N}$

$$z_j(t) = \int_{[t_0,t]} da_{N_j,\tau}(s) + y_j(t),$$

we rewrite the above as

$$\lim_{j\to+\infty} \|z_j - x_i\|_{\infty,[S+\delta_i,T]} = 0. \tag{3.18}$$

From inspection of (3.13) and (3.18), we have for all $j \in \mathbb{N}$ sufficiently large and $t \in [S+\delta_i, T]$ that

$$z_j(t) + (1/2)\rho\mathbb{B} \subset \text{int } A \quad \text{and} \quad z_j(t) \in \bar{x}(t) + (1/2)\varepsilon\mathbb{B}.$$

Denote by $\{s_1, ..., s_{M_j}\}$ the points of discontinuity of z_j in $[S+\delta_i, T]$. Notice that $\{s_1, ..., s_{M_j}\} \subset]S+\delta_i, T[$ since $S+\delta_i - \tau$ and $T-\tau$ are irrational. Moreover, for all $k \in \{1, ..., M_j\}$, $z_j : [t_{k-1}, t_k] \to \mathbb{R}^n$ is piece-wise absolutely continuous with at most one point of discontinuity at $s_k \in]t_{k-1}, t_k[$ where $t_k := S+\delta_i+\frac{k}{N}$, for all $k = 0, 1, ..., M_j - 1$, and $t_{M_j} := T$. We have

$$\begin{aligned} \phi\left(z_j(T)\right) - v\left(S+\delta_i, z_j\left(S+\delta_i\right)\right) &= v\left(T, z_j(T)\right) - v\left(S+\delta_i, z_j\left(S+\delta_i\right)\right) \\ &= \sum_{k=1}^{M_j} \left(v\left(t_k, z_j\left(t_k\right)\right) - v\left(t_{k-1}, z_j\left(t_{k-1}\right)\right)\right). \end{aligned}$$

Recalling (A.7) and (3.14), by Lemma 3.2.5 we conclude

$$\begin{aligned}
&\phi(z_j(T)) - v(S+\delta_i, z_j(S+\delta_i)) \\
&\geq -\sum_{k=1}^{M_j} \zeta^{\varepsilon}\left(s_k, z_j\left(s_k\right) - z_j\left(s_k^{-}\right)\right) - \int_{S+\delta_i}^{T} \tilde{\zeta}^{\varepsilon}(t, b(t), c(t))dt \\
&= -\frac{1}{N_j}\sum_{k=1}^{M_j} \zeta^{\varepsilon}\left(s_k, a\left(s_k\right)\right) - \int_{S+\delta_i}^{T} \tilde{\zeta}^{\varepsilon}(t, b(t), c(t))dt \\
&= -\int_{[S+\delta_i, T]}^{T} d\vartheta^{\varepsilon}_{N_j,\tau}(s) - \int_{S+\delta_i}^{T} \tilde{\zeta}^{\varepsilon}(t, b(t), c(t))dt.
\end{aligned}$$

Passing to the limit as $j \to +\infty$ and taking into account (3.17) and (3.18) we get

$$\begin{aligned}
&\phi\left(x_i(T)\right) - v\left(S+\delta_i, x_i\left(S+\delta_i\right)\right) \\
&\geq -\int_{S+\delta_i}^{T} \zeta^{\varepsilon}(t, a(t))dt - \int_{S+\delta_i}^{T} \tilde{\zeta}^{\varepsilon}(t, b(t), c(t))dt.
\end{aligned}$$

From (3.12) and the continuity of $v(.,.)$ on $[S, T] \times A$, we obtain from this last relation, in the limit as $i \to +\infty$,

$$\phi(x(T)) - v(S, x(S)) + \int_S^T \zeta^{\varepsilon}(t, a(t))dt + \int_S^T \tilde{\zeta}^{\varepsilon}(t, b(t), c(t))dt \geq 0.$$

But, since $\bar{x}(.)$ is a minimizer and $\zeta^{\varepsilon}(t, 0) = 0$, $\tilde{\zeta}^{\varepsilon}(t, 0, 0) = 0$, notice that

$$\begin{aligned}
&\phi(\bar{x}(T)) - v(S, \bar{x}(S)) \\
&+ \int_S^T \zeta^{\varepsilon}(t, \bar{a}(t) = 0)dt + \int_S^T \tilde{\zeta}^{\varepsilon}(t, \bar{b}(t) = 0, \bar{c}(t) = 0)dt = 0,
\end{aligned}$$

so $(\bar{x}(.), \bar{a}(.) \equiv 0, \bar{b}(.) \equiv 0, \bar{c}(.) \equiv 0)$ is a minimizer for (3.7). This ends the proof of Proposition 3.2.2.

3.3 Necessary Conditions Under State Constraints

We now focus on to the derivation of necessary optimality conditions. These conditions provide crucial insights into the structure of optimal solutions and serve as the foundation for developing numerical algorithms. We present the complete characterization of necessary conditions for infinite horizon optimal control problems subject to state constraints, extending classical results to the infinite time setting.

Assumption 3.3.1 Assumption 2.3.5 holds true with $F(t, x) = \{f(t, x, u) \mid u \in U(t)\}$ and $\Omega = A$.

Theorem 3.3.2 (Necessary conditions) *Consider Assumptions 3.1.1 and 3.3.1 and assume that $V(i,.)$ is locally Lipschitz continuous on A for all large $i \in \mathbb{N}$. Then V is locally Lipschitz continuous on $\mathbb{R}_0^+ \times A$.*

Moreover, if $(\bar{x},\bar{u})$ is optimal for $\mathscr{B}_\infty$ at $(t_0,x_0) \in (\mathbb{R}_0^+ \times \operatorname{int} A) \cap \operatorname{dom} V$, then there exist $p \in W^{1,1}_{\text{loc}}(\mathbb{R}^+_{t_0};\mathbb{R}^n)$, a nonnegative Borel measure μ on $\mathbb{R}^+_{t_0}$, and a Borel measurable function $\nu : \mathbb{R}^+_{t_0} \to \mathbb{R}^n$ such that, setting

$$q(t) = p(t) + \eta(t)$$

with

$$\eta(t) = \begin{cases} \int_{[t_0,t]} \nu(s)\, d\mu(s) & t \in]t_0,+\infty[, \\ 0 & t = t_0, \end{cases}$$

the following holds true:

(i) $\nu(t) \in \operatorname{cl}\operatorname{co} N_A(\bar{x}(t)) \cap \mathbb{B}$ *for* $\mu-$ a.e. $t \in \mathbb{R}^+_{t_0}$;

(ii) $p'(t) \in \operatorname{co}\left\{r \mid (r,q(t),-1) \in N_{\operatorname{graph} F(t,.)}(\bar{x}(t),\bar{x}'(t),L(t,\bar{x}(t),\bar{u}(t)))\right\}$ *for a.e.* $t \in \mathbb{R}^+_{t_0}$ *where*

$$F(t,x) := \{(f(t,x,u),L(t,x,u)) \mid u \in U(t)\};$$

(iii) $-p(t_0) \in \partial_x^+ V(t_0,\bar{x}(t_0))$, $-q(t) \in \widetilde{\partial}_x V(t,\bar{x}(t))$ *for a.e.* $t \in]t_0,+\infty[$;

(iv) for a.e. $t \in \mathbb{R}^+_{t_0}$

$$\begin{aligned} &\langle q(t), f(t,\bar{x}(t),\bar{u}(t))\rangle - L(t,\bar{x}(t),\bar{u}(t)) \\ &= \max_{u\in U(t)} \langle q(t), f(t,\bar{x}(t),u)\rangle - L(t,\bar{x}(t),u); \end{aligned}$$

(v) $(H(t,\bar{x}(t),q(t)),-q(t)) \in \widetilde{\partial} V(t,\bar{x}(t))$ *for a.e.* $t \in]t_0,+\infty[$;

where

$$\begin{aligned} \widetilde{\partial}_x V(t,x) &:= \operatorname{Lim\,sup}_{y \underset{\operatorname{int} A}{\to} x} \operatorname{co} \partial_x V(t,y), \\ \widetilde{\partial} V(t,x) &:= \operatorname{Lim\,sup}_{(s,y) \underset{\mathbb{R}_0^+ \times \operatorname{int} A}{\longrightarrow} (t,x)} \operatorname{co} \partial V(s,y). \end{aligned} \tag{3.19}$$

Remark 3.3.3 (i) Denote $\mathscr{F}(t,x) := \{f(t,x,u) | u \in U(t)\}$ for any t,x. By employing a similar argument as the one used in the proof of Theorem 3.3.2, we can conclude that the same result holds true when we assume Assumption 3.1.1:*(I)–(III)* with the additional:

- for any $R > 0$ there exists $r > 0$ such that $\mathscr{F}(.,x)$ is absolutely continuous from the left, uniformly over $x \in (\operatorname{bdr} A + r\mathbb{B}) \cap B(0,R)$;
- (Relaxed Inward Pointing Condition) for any $(t,x) \in \mathbb{R}_0^+ \times \operatorname{bdr} A$

$$\operatorname{Lim\,inf}_{\substack{(s,y)\to(t,x)\\ (s,y)\in\mathbb{R}_0^+\times A}} \operatorname{co}\mathscr{F}(s,y) \bigcap \operatorname{int} T_A^C(x) \neq \emptyset.$$

We note that, if $\mathscr{F}$ is continuous, then the inward pointing condition in Assumption 3.3.1 condition reduces to

$$\operatorname{co}\mathscr{F}(t,x)\bigcap \operatorname{int} T_A^C(x) \neq \emptyset \qquad \forall (t,x) \in \mathbb{R}_0^+ \times \operatorname{bdr} A. \tag{3.20}$$

In that case, if $\mathscr{H}$ denotes the Hamiltonian

$$\mathscr{H}(t,x,p) = \max_{v\in\mathscr{F}(t,x)} \langle p, v\rangle \qquad \forall\ (t,x,p) \in \mathbb{R}_0^+ \times \mathbb{R}^n \times \mathbb{R}^n,$$

then, by the separation theorem, (3.20) is equivalent to

$$\mathscr{H}(t,x,-p) > 0 \qquad \forall\, 0 \neq p \in N_A^C(x).$$

(ii) Thanks to Remark 3.3.3:*(i)* and to (3.20), if f is time indpendent and $U(.) \equiv U$, the inward pointing condition in (3.20) can be replaced by the simpler condition

$$\max_{u\in U} \langle -p, f(x,u)\rangle > 0 \qquad \forall p \in N_A^C(x)\backslash\{0\},\ \forall x \in \operatorname{bdr} A.$$

(iii) Consider F as in the statement of Theorem 3.3.2 and assume that for all $R > 0$ there exists $r > 0$ such that $F(.,x)$ is absolutely continuous from the left uniformly over $(\operatorname{bdr} A + r\mathbb{B}) \cap B(0,R)$. Then the conclusion of Theorem 3.3.2 holds if Assumption 3.3.1 is replaced by the relaxed inward pointing condition in Remark 3.3.3:*(i)*.

(iv) Theorem 3.3.2 implies a weaker hamiltonian inclusion

$$(-p'(t), \bar{x}'(t)) \in \operatorname{co}\partial_{(x,p)} H(t, \bar{x}(t), q(t)) \qquad \text{a.e. } t \in \mathbb{R}_{t_0}^+.$$

(v) If $V(i,.)$ is locally Lipschitz continuous on A for all large i, then, under assumptions of Theorem 3.3.2, $V(t,.)$ is locally Lipschitz on A for every $t \geq 0$.

(vi) Sufficient conditions for the Lipschitz continuity of $V(t,.)$ in the nonautonomous case for unbounded A are investigated in Chap. 2.

(vii) Theorem 3.3.2 provide the normal maximum principle (i.e. $q_0 = 1$) together with partial and full sensitivity relations and a transversality condition at the initial time, under mild assumption on dynamics and constraints. To describe the results in the smooth case, assume for the sake of simplicity that $L(t,x,u) = e^{-\lambda t} l(x,u)$ is smooth, $U(.) \equiv U$ is a closed subset of $\mathbb{R}^m$, $V(t,.)$ is continuously differentiable, and denote by $N_A(y)$ the limiting normal cone to A at y. If $(\bar{x},\bar{u})$ is optimal for $\mathscr{B}_\infty$ at $(t_0,x_0) \in \mathbb{R}_0^+ \times \operatorname{int} A$, then Theorem 3.3.2 below guarantees the existence of a locally absolutely continuous co-state $p(.)$, a nonnegative Borel measure μ on $\mathbb{R}_{t_0}^+$, and a Borel measurable selection $\nu(.) \in \operatorname{cl}\operatorname{co} N_A(\bar{x}(.)) \cap \mathbb{B}$ such that $p(.)$ satisfies, for a.e. $t \in \mathbb{R}_{t_0}^+$, the adjoint equation

$$-p'(t) = \mathrm{d}_x f(t, \bar{x}(t), \bar{u}(t))^\top (p(t) + \eta(t)) - e^{-\lambda t}\nabla_x l(\bar{x}(t), \bar{u}(t)),$$

the maximality condition

$$\begin{aligned}&\langle\, p(t) + \eta(t), f(t, \bar{x}(t), \bar{u}(t))\rangle - e^{-\lambda t} l(\bar{x}(t), \bar{u}(t))\\ &= \max_{u\in U} \left\{\langle\, p(t) + \eta(t), f(t, \bar{x}(t), u)\rangle - e^{-\lambda t} l(\bar{x}(t), u)\right\},\end{aligned}$$

and, for a.e. $t \in]t_0, +\infty[$, the transversality and sensitivity relations

$$-p(t_0) = \nabla_x V(t_0, \bar{x}(t_0)), \quad -(p(t) + \eta(t)) = \nabla_x V(t, \bar{x}(t)), \tag{3.21}$$

where $\eta(t_0) = 0$ and $\eta(t) = \int_{[t_0,t]} \nu(s)\, d\mu(s)$ for all $t \in]t_0, +\infty[$. Observe that, if $\bar{x}(.) \in \text{int}\, A$, then $\nu(.) \equiv 0$ and the usual maximum principle holds true. But if $\bar{x}(t) \in \text{bdr}\ A$ for some time t, then a measure multiplier factor, $\int_{[0,t]} \nu\, d\mu$, may arise modifying the adjoint equation.

(viii) From Theorems 2.5.7 and 3.3.2:*(iii)*, whenever $V(t, .) = e^{-\lambda t} v(.)$, it follows that

$$\lim_{t\to+\infty} q(t) = 0.$$

Proof of Theorem 3.3.2 For any $j \in \mathbb{N}$ such that $j \geq t_0$ consider the Bolza problem $\mathscr{B}_j$. We can rewrite such problem as a Mayer one on $\mathbb{R}^{n+1}$: let $\mathscr{M}(\phi^j, j)$ be the Mayer problem on $\mathbb{R}^{n+1}$

$$\begin{cases}\text{minimize } \phi^j(x(j), z(j)) \text{ over all} & \\ (x', z')(t) \in \overline{F}(t, x(t), z(t)) & \text{a.e. } t \in [t_0, j],\\ (x, z) \in W^{1,1}([t_0, j]; \mathbb{R}^n) \times W^{1,1}([t_0, j]; \mathbb{R}), & \\ x(t_0) = x_0,\, z(t_0) = 0, & \\ (x, z)(t) \in \Omega & \forall t \in [t_0, \tau],\end{cases} \tag{3.22}$$

where we define for all $t \in \mathbb{R}_0^+$ and all $(x, z) \in \mathbb{R}^n \times \mathbb{R}$

$$\begin{aligned}\phi^j(x, z) &:= V(j, x) + z,\\ \overline{F}(t, x, z) &:= F(t, x),\\ \Omega &:= A \times \mathbb{R}.\end{aligned}$$

Denoting by V^j the extended value function on $[0, j] \times \Omega$ for problem $\mathscr{M}(\phi^j, j)$ it follows, by standard arguments, that

$$V^j(t, x, z) = V_{\mathscr{B}_j}(t, x) + z \tag{3.23}$$

for all $(t, x, z) \in [0, j] \times \Omega$. Since, for all large j, $V(j, .)$ is locally Lipschitz continuous on A, also $\phi^j(., .)$ is locally Lipschitz on Ω. For every j, by standard analysis arguments, we can extend $\phi^j(., .)$ (we keep the same notation) to a Lipschitz continuous function on whole $\mathbb{R}^{n+1}$ without changing the value function of the Bolza problem $\mathscr{B}_j$. From Lemma 3.2.3, it follows that V^j is locally Lipschitz on $[0, j] \times \Omega$ for all large j. Then $V_{\mathscr{B}_j}$ is locally Lipschitz on $[0, j] \times A$ and so, by Lemma 3.2.1, the value function V is locally Lipschitz on $[0, j] \times A$. By the arbitrariness of j, V is locally Lipschitz continuous on $\mathbb{R}_0^+ \times A$. Hence, if $T > 0$, from (3.23) and (3.3) it follows that V^j's are uniformly locally Lipschitz continuous on $[0, T] \times \Omega$ for all $j \geq T$.

Consider $(\bar{x}, \bar{u})$ (optimal) at $(t_0, x_0) \in (\mathbb{R}_{t_0}^+ \times \text{int } A) \cap \text{dom } V$. Let $j \in \mathbb{N} \cap \mathbb{R}_{t_0}^+$. Notice that, since $V(., .)$ is l.s.c. (see Proposition 4.2.4:*(ii)* for a proof), an optimal trajectory exists at (t_0, x_0).

We dive the proof of Theorem 3.3.2 into four steps.

STEP 1: FINITE HORIZON APPROXIMATIONS AND DISTANCE ESTIMATES. We first deduce the main implications of necessary conditions on finite horizon, i.e. Theorem A.1.1, in presence of set state constraints for the Mayer problem (3.22). Before anything else, we rewrite the problem (3.22) with set state constraints as in functional constrains fashion. From now on, let we formally denote the variable

$$\mathrm{x} = (x, z)$$

and consider a minimizer $\bar{\mathrm{x}}(.)$ for (3.22). By applying Propostion 3.2.2, we have that for every $\varepsilon > 0$ sufficiently small there exists $\delta > 0$ such that

$$(\bar{\mathrm{x}}, \bar{X} := (\bar{a} \equiv 0, \bar{b} \equiv 0),\ \bar{Y} := (\bar{c} \equiv 0, \bar{d} \equiv 0, \bar{y} \equiv V^{\text{ext}}(\bar{\mathrm{x}}(t_0))))$$

is a δ-minimizer, for the optimal control problem

$$\text{minimize}\phi^j(\mathrm{x}(j)) + y(j) + d(j)$$

over all

$$\begin{aligned} (\mathrm{x}, X := (a, b)) &\in (W^{1,1}([t_0, j]; \mathbb{R}^k))^3, \\ Y := (c, d, y) &\in (W^{1,1}([t_0, j]; \mathbb{R}))^3, \end{aligned}$$

satisfying

$$\begin{cases} (\mathrm{x}'(t), X'(t), Y'(t)) \in \mathcal{F}(t, \mathrm{x}(t), X(t), Y(t)) & \text{a.e. } t \in [t_0, j], \\ (\mathrm{x}(t), X(t)) \in \Omega \times \left(\mathbb{R}^k\right)^2 & \forall t \in [t_0, j], \\ Y(t) \in \mathbb{R}^3 & \forall t \in [t_0, j], \\ y(t_0) \geq V^{\text{ext}}(\mathrm{x}(t_0)), \\ (a(t_0), b(t_0), c(t_0), d(t_0)) = 0, \end{cases}$$

where $V^{\text{ext}} : \mathbb{R}^k \to]-\infty, +\infty]$ is defined by

$$V^{\text{ext}}(\mathrm{x}) := \begin{cases} -V(t_0, \mathrm{x}) & \text{if } \mathrm{x} \in \Omega, \\ +\infty & \text{otherwise,} \end{cases}$$

$\mathcal{F} : [t_0, j] \times (\mathbb{R}^n)^3 \times \mathbb{R}^3 \rightsquigarrow (\mathbb{R}^n)^3 \times \mathbb{R}^3$ is the set-valued map

$$\begin{aligned} &\mathcal{F}(t, \mathrm{x}, (a, b), (c, d, y)) \\ &:= \{(a + (1 + c)(b + v), e_1, e_2, e_3, \zeta^{\varepsilon}(t, a) + \tilde{\zeta}^{\varepsilon}(t, b, c), 0) \\ &\qquad \text{such that } v \in \overline{F}(t, \mathrm{x}), e_1, e_2 \in \varepsilon\mathbb{B}, e_3 \in [-\varepsilon, +\varepsilon]\}, \end{aligned}$$

and, with a slight abuse of notation, $V(t, \mathrm{x})$ denotes the value function of the Mayer problem (3.22). Applying Theorem 2.3.6, and replacing $\delta > 0$ with a suitable small one, there exists a constant $C > 0$ (depending only on the time interval $[t_0, j]$) such that the trajectory $(\bar{\mathrm{x}}(.), \bar{X}(.), \bar{Y}(.), \bar{z} \equiv 0)$ is still a δ-minimizer for the problem

$$\text{minimize } \phi^j(\mathrm{x}(j)) + y(j) + d(j) + Ce(j),$$

over all $(\mathrm{x}, X) \in \left(W^{1,1}\left([t_0, j]; \mathbb{R}^k\right)\right)^3$, $(Y, e) \in \left(W^{1,1}([t_0, j]; \mathbb{R})\right)^4$ such that

$$\begin{cases} (\mathrm{x}'(t), X'(t), Y'(t), e'(t)) \in \mathcal{F}(t, \mathrm{x}(t), X(t), Y(t)) \times \{0\} & \text{a.e. } t \in [t_0, j], \\ (\mathrm{x}(t), X(t)) \in \mathbb{R}^n \times \left(\mathbb{R}^k\right)^2 & \forall t \in [t_0, j], \\ (Y(t), e(t)) \in \mathbb{R}^4 & \forall t \in [t_0, j], \\ y(t_0) \geq V^{\text{ext}}(\mathrm{x}(t_0)) & \\ d_\Omega(\mathrm{x}(t)) - e(t) \leq 0 & \forall t \in [t_0, j], \\ (a(t_0), b(t_0), c(t_0), d(t_0)) = 0. & \end{cases}$$

We can now apply Theorem A.1.1. Letting $Z := (X, Y, e)$, the state constraint is

$$g(\mathrm{x}, \mathrm{Z}) := d_\Omega(\mathrm{x}) - e \leq 0$$

and involving the sub-differential

$$\begin{aligned} \partial_{\#} g(\bar{\mathrm{x}}(t), \bar{Z}) = \mathrm{cl\ co}\{\xi \mid \exists \xi_i \to \xi, \exists\, (\mathrm{x}_i, Z_i) \to (\bar{\mathrm{x}}(t), \bar{Z}(t)) \text{ such that} \\ \xi_i = \nabla g\,(\mathrm{x}_i, Z_i) \text{ and } g(\mathrm{x}_i, Z_i) > 0 \text{ for each } i\}. \end{aligned}$$

From the relations of limiting subdifferentials to the distance function (see Sect. 1.1) and standard subdifferential calculus, we know

$$\partial_{\#} g(\bar{x}(t), \bar{Z}(t)) \subset \begin{cases} (\text{cl co } N_{\Omega}(\bar{x}(t))) \cap \mathbb{B} \times \{(0,0,0,0,0)\} \times \{-1\} & \text{if } \bar{x}(t) \in \text{bdr } \Omega, \\ (0,0,0,0,0,0,-1) & \text{if } \bar{x}(t) \in \text{int } \Omega. \end{cases}$$

Hence, we conclude that there exist:

(I) a co-state trajectory

$$\hat{P}(.) :=(P(.), p_a(.), p_b(.), p_c(.), p_d(.), p_y(.), p_z(.)) \in W^{1,1}([t_0, j]; \mathbb{R}^{3k+4});$$

(II) a number $\lambda \geq 0$;
(III) a nonnegative Borel measure μ on $[t_0, j]$ with

$$\text{supp } \mu \subset \{t \mid \bar{x}(t) \in \text{bdr } \Omega\};$$

(IV) a Borel measurable selection

$$\Pi(t) \in \text{cl co } N_{\Omega}(\bar{x}(t)) \cap \mathbb{B} \quad \mu - \text{a.e. } t \in [t_0, j];$$

satisfying

$$(\hat{P}(.), \mu, \lambda) \neq 0$$

and:

(A) $\hat{P}'(t) \in \text{co}\left\{\alpha \mid (\alpha, \hat{Q}(t)) \in N_{\text{graph } \mathcal{F}(t,.)}((\bar{x}(t), \bar{Z}(t)), (\bar{x}'(t), \bar{Z}'(t)))\right\}$ a.e. t;
(B) $-\hat{Q}(j) \in \lambda \partial \phi^j(\bar{x}(j)) \times \{0\} \times \{0\} \times \{0\} \times \{\lambda\} \times \{\lambda\} \times \{\lambda C\}$;
(C) $(P(t_0), p_y(t_0)) \in N_{\text{epi } V^{\text{ext}}}(\bar{x}(t_0), V^{\text{ext}}(\bar{x}(t_0)))$ and $p_z(t_0) = 0$;
(D) $\langle \hat{Q}(t), (\bar{x}'(t), \bar{Z}'(t)) \rangle \geq \langle \hat{Q}(t), v \rangle$ for a.e. t and for all $v \in \mathcal{F}(t, \bar{x}(t), \bar{Z}(t))$;

where $\hat{Q} : [t_0, j] \to \mathbb{R}^{3k+4}$ is the function defined by $\hat{Q}(t_0) = \hat{P}(t_0)$ and for all $t \in]t_0, j]$

$$\hat{Q}(t) := \hat{P}(t) + \left(\int_{[t_0,t]} \Pi(s) d\mu(s), 0, 0, 0, 0, 0, - \int_{[t_0,t]} d\mu(s) \right).$$

The function $t \mapsto \int_{[t_0,t[} \Pi(s) d\mu(s)$ is of bounded variation and it differs from $t \mapsto \int_{[t_0,t]} \Pi(s) d\mu(s)$ on a set of measure zero.

Using Lemma 1.1.1 we deduce that, if

$$\mathbb{R}^7 \times \mathbb{R}^7 \ni (\alpha, \beta) \in N_{\text{graph } \mathcal{F}(t,.)}((\bar{x}(t), \bar{Z}(t)), (\bar{x}'(t), \bar{Z}'(t)))$$

then:

(P1) $\alpha_5 = \alpha_6 = \alpha_7 = 0$ and $\beta_2 = 0, \beta_3 = \beta_4 = 0$;

(P2) $(\alpha_1, \beta_1) \in N_{\text{graph } \bar{F}(t,.)}(\bar{\mathrm{x}}(t), \bar{\mathrm{x}}'(t))$ and $-(\alpha_2 + \beta_1) \in \beta_5 \operatorname{co} \partial_a \zeta^\varepsilon(t, \bar{a}(t) = 0)$;
(P3) $-\big(\alpha_3 + \beta_1, \alpha_4 + \langle \beta_1, \bar{\mathrm{x}}'(t)\rangle\big) \in \beta_5 \operatorname{co} \partial_{(b,c)} \tilde{\zeta}^\varepsilon(t, \bar{b}(t) = 0, \bar{c}(t) = 0)$.

So, from (A) we have that

$$p_a(.) \equiv 0,\ p_b(.) \equiv 0,\ p_c(.) \equiv 0 \text{ and } p_d'(.) \equiv 0,\ p_y'(.) \equiv 0,\ p_z'(.) \equiv 0, \tag{3.24}$$

and

$$\begin{cases} P'(t) \in \operatorname{co}\left\{r \mid (r, Q(t)) \in N_{\text{graph } \bar{F}(t,.)}(\bar{\mathrm{x}}(t), \bar{\mathrm{x}}'(t))\right\} \text{ a.e.} t, \\ \text{with } Q(t) := P(t) + \int_{[t_0,t]} \Pi(s) d\mu(s) \text{ for } t \in]t_0, j]. \end{cases} \tag{3.25}$$

From (B) it follows that

$$-Q(j) \in \lambda \partial \phi^j(\bar{\mathrm{x}}(j))$$

and

$$p_d(j) = -\lambda,\ p_y(j) = -\lambda,\ p_z(j) - \int_{[t_0,j]} d\mu(s) = -\lambda C. \tag{3.26}$$

Since $p_z(t_0) = 0$, we know $p_z(.) \equiv 0$ and then $\int_{[t_0,t]} d\mu(s) = \lambda C$. We can deduce that $\lambda > 0$. Indeed, conversely, by the preceding equation $\mu = 0$. Recalling (3.25) and Lemma 1.3.2, this leads to $P(.) = 0$. A contradiction since $(P(.), \mu, \lambda) \neq 0$. We can therefore assume, by scaling the multipliers, that $\lambda = 1$ and so $p_y(t_0) = -1$. We have

$$(P(t_0), -1) \in N_{\text{epi } V^{\text{ext}}}(\bar{\mathrm{x}}(t_0), V^{\text{ext}}(\bar{\mathrm{x}}(t_0))),$$

hence

$$P(t_0) \in \partial V^{\text{ext}}(\bar{\mathrm{x}}(t_0)).$$

Notice also that (D) implies that for a.e. t,

$$\langle Q(t), \bar{\mathrm{x}}'(t)\rangle \geq \langle Q(t), v\rangle \quad \forall v \in \bar{F}(t, \bar{\mathrm{x}}(t)).$$

We deduce from (3.24) that $p_a'(.) \equiv 0$, $p_b'(.) \equiv 0$, $p_c'(.) \equiv 0$. Since $p_d(.) \equiv -1$, it follows from (P2)–(P3) and taking into account $\partial \sigma_X(y) = \arg\max_{\mathrm{x} \in \mathrm{cl}(\mathrm{co}\ X)} \langle y, \mathrm{x}\rangle$ (here $\sigma_X(.)$ stands for the support function of the set X, i.e. $\sigma_X(\xi) := \sup_{\mathrm{x} \in X} \langle \mathrm{x}, \xi\rangle$),

$$Q(t) \in \operatorname{co} \partial_a \zeta^\varepsilon(t, \bar{a}(t) = 0) \subset \{\xi \in \operatorname{co} \partial_{\mathrm{x}} V(t, \mathrm{x}) \mid \mathrm{x} \in (\bar{\mathrm{x}}(t) + \varepsilon \mathbb{B}) \cap \operatorname{int} \Omega\}, \tag{3.27}$$

and

$$\begin{aligned} &(Q(t), \langle Q(t), \bar{\mathrm{x}}'(t)\rangle) \\ &\in \operatorname{co} \partial_{(b,c)} \tilde{\zeta}^\varepsilon(t, \bar{b}(t) = 0, \bar{c}(t) = 0) \\ &\subset \{(\xi, \eta) \in \operatorname{co} \partial V(t, \mathrm{x}) \mid (t, \mathrm{x}) \in ((t, \bar{\mathrm{x}}(t)) + \varepsilon \mathbb{B}) \cap ([t_0, j] \times \operatorname{int} \Omega)\}. \end{aligned} \tag{3.28}$$

Now, let $\varepsilon_i \to 0+$. For any i, from the preceding analysis with $\varepsilon = \varepsilon_i$, we obtain multipliers, which we label $P_i(.)$, $\Pi_i(.)$, $\mu_i(.)$, and $Q_i(.)$. It can be deduced from (3.27) and the local Lipschitz continuity of V that $(Q_i(.))_{i\in\mathbb{N}}$ is bounded in $L^\infty\left([t_0, j]; \mathbb{R}^k\right)$. We infer from Lemma 1.3.2, (2.71), and (3.25), that there exists $R = R(j - t_0) > 0$ such that

$$|P_i'(t)| \leq h_R(t)\,|Q_i(t)| \ \text{ for a.e. } t \in [t_0, j], \ \text{ with } h_R(t) := \alpha_R(t) + c_R(t).$$

Thus, for some $\ell \in L^1([t_0, j]; \mathbb{R}_0^+)$ and all large $i \in \mathbb{N}$, $\left|p_i'(t)\right| \leq \ell(t)$ a.e. t. Since $-Q_i(j) = -P_i(j) - \int_{[t_0,j]} d\mu_i(s) \in \partial\phi^j(\bar{\mathrm{x}}(j))$ and, from (3.26), $\int_{[t_0,j]} d\mu_i(s) = C$, then also $(P_i(j))_{i\in\mathbb{N}}$ is bounded. The sequence $(P_i(.))_{i\in\mathbb{N}}$ is so bounded in $L^\infty\left([t_0, j]; \mathbb{R}^k\right)$. Now, define the measure $\Phi_i \in C\left([t_0, j]; \mathbb{R}^k\right)^*$ by $\Phi_i(dt) = \Pi_i(t)\mu_i(dt)$. Extracting sub-sequences and taking the same notation, we can conclude that, for some absolutely continuous function $P : [t_0, j] \to \mathbb{R}^k$, some positive Borel measure μ on $[t_0, j]$, and functions of bounded variation $\Phi(.)$ and $Q(.)$: $P_i \to P$ in $C([t_0, j]; \mathbb{R}^k)$; $P_i' \rightharpoonup P'$ in $L^1\left([t_0, j]; \mathbb{R}^k\right)$; $\mu_i \rightharpoonup^* \mu$ in $C([t_0, j]; \mathbb{R})^*$; $\Phi_i \rightharpoonup^* \Phi$ in $C\left([t_0, j]; \mathbb{R}^k\right)^*$; and $Q_i(t) \to Q(t)$ a.e. t.

Lemma 3.3.4 *Let $(\mu_i)_i$ be a sequence of nonnegative Borel measures on $[S, T]$ weakly* convergent to μ_0 and $(\Pi_i)_i : [S, T] \to \mathbb{R}^n$ be a sequence of Borel measurable functions and uniformly bounded. Assume that for a suitable $R > 0$ and for all $t \in [S, T]$: $A(t)$ is a nonempty closed convex subset of $\mathbb{R}^n$ and $A(t) \subset R\mathbb{B}$. Assume also that $\Pi_i(t) \in A(t)$ for μ_i-a.e. t and all $i \in \mathbb{N}$. Then there exists a Borel measurable function $\Pi_0 : [S, T] \to \mathbb{R}^n$ such that, along a subsequence, the measures $\Phi_i(dt) := \Pi_i(t)\mu_i(dt)$ converge weakly* to the measure $\Pi_0(t)\mu_0(dt)$ and $\Pi_0(t) \in A(t)$ for μ_0-a.e. t.*

Proof See Sect. A.3.2. □

From the above lemma there exists a Borel measurable selection

$$\Pi(t) \in (\text{cl co}\, N_\Omega(\bar{\mathrm{x}}(t))) \cap \mathbb{B} \text{ such that } \Phi(dt) = \Pi(t)\mu(dt).$$

Moreover, using a standard limiting argument, the above assertions (A)–(D) are also satisfied for the limiting multipliers P, Q, μ, Π, Φ. Furthermore, by inspection of (3.27) and (3.28), and the definitions (3.19) we deduce

$$-\,Q(t) \in \widetilde{\partial}_{\mathrm{x}} V(t, \bar{\mathrm{x}}(t)) \quad \text{a.e. } t \in]t_0, j[,$$

$$(\tilde{H}(t, \bar{\mathrm{x}}(t), Q(t)), -Q(t)) \in \widetilde{\partial} V(t, \bar{\mathrm{x}}(t)) \quad \text{a.e. } t \in]t_0, j[,$$

with $\tilde{H}(t, \mathrm{x}, P) = \sup_{v\in \tilde{F}(t,\mathrm{x})} \langle P, v\rangle$.

STEP 2: FROM MAYER TO BOLZA. Since the restriction of $(\bar{x}, \bar{u})$ to $[t_0, j]$ is optimal for $V_{\mathscr{B}_j}$ at (t_0, x_0), setting

$$\bar{z}(t) = \int_{t_0}^{t} L(s, \bar{x}(s), \bar{u}(s))ds,$$

we have that the restriction of $(\bar{\mathrm{x}} := (\bar{x}, \bar{z}), \bar{u})$ to $[t_0, j]$ is optimal for V^j at $(t_0, (x_0, 0))$ too. From Step 1, there exist absolutely continuous arcs $(P_j)_j$ and functions $(\Phi_j)_j$ of bounded variation defined on $[t_0, j]$, and nonnegative measures $(\mu_j)_j$ on $[t_0, j]$ such that $(\Phi_j)_j$ are continuous from the right on $]t_0, j[$ and:

(a) $Q_j(t) = P_j(t) + \Phi_j(t)$, where $\Phi_j(t_0) = 0$, $\Phi_j(t) = \int_{[t_0,t]} \Pi_j(s)\, d\mu_j(s)$ for all $t \in]t_0, j]$ for some Borel measurable selections $\Pi_j(s) \in \mathrm{cl\,co}\, N_\Omega(\bar{X}(s)) \cap B$ μ_j–a.e. $s \in [t_0, j]$;
(b) $P_j'(t) \in \mathrm{co}\left\{R \mid (R, Q_j(t)) \in N_{\mathrm{graph}\, \overline{F}(t,.)}(\bar{\mathrm{x}}(t), \bar{\mathrm{x}}'(t))\right\}$ for a.e. $t \in [t_0, j]$;
(c) $-Q_j(t_0) \in \partial_{\mathrm{x}}^{+} V^j(t_0, \bar{\mathrm{x}}(t_0))$;
(d) $\langle Q_j(t), \bar{\mathrm{x}}'(t)\rangle = \max\left\{\langle Q_j(t), v\rangle \mid v \in \overline{F}(t, \bar{\mathrm{x}}(t))\right\}$ for a.e. $t \in [t_0, j]$;
(e) $-Q_j(t) \in \widetilde{\partial}_{\mathrm{x}} V^j(t, \bar{\mathrm{x}}(t))$ for a.e. $t \in]t_0, j]$;
(f) $(\widetilde{H}(t, \bar{\mathrm{x}}(t), Q_j(t)), -Q_j(t)) \in \widetilde{\partial} V^j(t, \bar{\mathrm{x}}(t))$ for a.e. $t \in]t_0, j]$.

Let $P_j(t) = (p_j(t), p_j^0(t))$, $Q_j(t) = (q_j(t), q_j^0(t))$, $\Phi_j(t) = (\eta_j(t), \eta_j^0(t))$, and $\Pi_j(t) = (\nu_j(t), \nu_j^0(t))$. Using the definition of limiting normal vectors as limits of strict normal vectors, relations (a)–(c), and the fact that $N_\Omega(\bar{\mathrm{x}}(.)) = N_A(\bar{x}(.)) \times \{0\}$ we obtain for a.e. $t \in [t_0, j]$

$$p_j'(t) \in \mathrm{co}\left\{r \mid (r, q_j(t), q_j^0(t)) \in N_{\mathrm{graph}\, F(t,.)}(\bar{x}(t), \bar{x}'(t), L(t, \bar{x}(t), \bar{u}(t)))\right\},$$

with

$$(p_j^0)' \equiv 0, \quad p_j^0(j) + \eta_j^0(j) = -1, \quad \nu_j^0 \equiv 0.$$

Thus, on account of (d)–(f), for a.e. $t \in [t_0, j]$ we derive the Extended Euler-Lagrange Condition

$$p_j'(t) \in \mathrm{co}\left\{r \mid (r, q_j(t), -1) \in N_{\mathrm{graph}\, F(t,.)}(\bar{x}(t), \bar{x}'(t), L(t, \bar{x}(t), \bar{u}(t)))\right\} \tag{3.29}$$

where $q_j(t) = p_j(t) + \eta_j(t)$, with

$$\eta_j(t) = \begin{cases} \int_{[t_0,t]} \nu_j(s)\, d\mu_j(s) & t \in]t_0, j], \\ 0 & t = t_0, \end{cases} \tag{3.30}$$

and $\nu_j(t) \in \mathrm{cl\,co}\, N_A(\bar{x}(t)) \cap \mathbb{B}$ for μ_j–a.e. on $[t_0, j]$, satisfy the maximum principle

$$\begin{aligned} &\langle q_j(t), f(t, \bar{x}(t), \bar{u}(t))\rangle - L(t, \bar{x}(t), \bar{u}(t)) \\ &= \max_{u \in U(t)} \langle q_j(t), f(t, \bar{x}(t), u)\rangle - L(t, \bar{x}(t), u) \qquad \text{a.e. } t \in [t_0, j], \end{aligned} \tag{3.31}$$

the transversality condition in terms of limiting superdifferential

$$-p_j(t_0) \in \partial_x^+ V(t_0, x_0), \tag{3.32}$$

and the sensitivity relations

$$-q_j(t) \in \widetilde{\partial}_x V(t, \bar{x}(t)) \qquad \text{a.e. } t \in]t_0, j], \tag{3.33}$$

$$(H(t, \bar{x}(t), q_j(t)), -q_j(t)) \in \widetilde{\partial} V(t, \bar{x}(t)) \qquad \text{a.e. } t \in]t_0, j]. \tag{3.34}$$

We extend the functions p_j and η_j to whole interval $]j, +\infty[$ as the constants $p_j(j)$ and $\eta_j(j)$, respectively. We denote again by p_j and η_j such extensions.

STEP 3: UNIFORM ESTIMATES. We need first the following.

Lemma 3.3.5 *Let $I \subset \mathbb{R}$ be a compact interval and $\Phi : I \rightsquigarrow \mathbb{R}^n$ be a lower semicontinuous set valued map such that $\Phi(t)$ is a closed convex cone and* int $\Phi(t) \neq \emptyset$ *for all $t \in I$. Then for every $\varepsilon > 0$ there exists a continuously differentiable function $f : \{t \in \mathbb{R} \mid d_I(t) < \varepsilon\} \to \mathbb{R}^n$ such that for all $t \in \{s \in I \mid \Phi(s) \neq \mathbb{R}^n\}$*

$$\sup_{q \in \Phi(t)^- \cap \mathbb{S}} \langle q, f(t) \rangle \leq -\varepsilon.$$

Proof Define the compact set $\mathscr{E} := \{t \in I \mid \Phi(t) \neq \mathbb{R}^n\} \subset \mathbb{R}$ and

$$\Gamma(t) := \Phi(t)^- \cap \mathbb{S} \quad \forall t \in \mathscr{E}.$$

It is sufficient to consider the case $\mathscr{E} \neq \emptyset$.

We claim that $\Gamma(.)$ is upper semicontinuous on $\mathscr{E}$. Indeed, $\Gamma(.)$ has nonempty compact images and is bounded, so it is sufficient to to show that graph Γ is closed. Let $q_i \in \Gamma(t_i) \subset \Phi(t_i)^-$, $q_i \to q$, $t_i \to_{\mathscr{E}} t_0$. Then $|q| = 1$. Consider $v \in \Phi(t_0)$. By the lower semicontinuity of $\Phi(.)$ there exist $v_i \in \Phi(t_i)$, $v_i \to v$. So, $\langle q_i, v_i \rangle \leq 0$ and, taking the limit, we have $\langle q, v \rangle \leq 0$. Since $v \in \Phi(t_0)$ is arbitrary, $q \in \Phi(t_0)^-$ and our claim is proved.

Now, fix $\varepsilon > 0$ and define $K_\varepsilon(t) := \{v \in \Phi(t) \mid \langle q, v \rangle \leq -2\varepsilon \text{ for all } q \in \Gamma(t)\}$ for all $t \in \mathscr{E}$. Since int $\Phi(t) \neq \emptyset$, also $K_\varepsilon(t) \neq \emptyset$ for all $t \in \mathscr{E}$. Moreover, $K_\varepsilon(.)$ has closed convex images and is lower semicontinuous on $\mathscr{E}$. So by Michael's Theorem there exists a continuous function $\bar{f} : \mathscr{E} \to \mathbb{R}^n$ such that $\bar{f}(t) \in K_\varepsilon(t) \subset \text{int } \Phi(t)$ for all $t \in \mathscr{E}$. Let $f^{\text{ext}} : \mathbb{R} \to \mathbb{R}^n$ be any continuous extension of $\bar{f}$. Then $f^{\text{ext}}(t) \in K_\varepsilon(t) \subset \text{int } \Phi(t)$ for any $t \in \mathscr{E}$. Consider now a continuously differentiable function $f : \{t \in \mathbb{R} \mid d_I(t) < \varepsilon\} \to \mathbb{R}^n$ such that $\|f - f^{\text{ext}}\|_{\infty, I} \leq \varepsilon$. Then for all $t \in \mathscr{E}$ and all $q \in \Phi(t)^- \cap \mathbb{S}$ we have $\langle q, f(t) \rangle = \langle q, f(t) - f^{\text{ext}}(t) \rangle + \langle q, f^{\text{ext}}(t) \rangle \leq \varepsilon - 2\varepsilon = -\varepsilon$, and the proof is complete. □

Lemma 3.3.6 *For all $j \in \mathbb{N}$ let $\phi^j : \mathbb{R}^n \to \mathbb{R}$ be a locally Lipschitz continuous function. Fix $(t_0, x_0) \in \mathbb{R}_0^+ \times \text{int } A$, $T > t_0$ and consider the problems $\mathscr{M}(\phi^j, j)$. Assume also that $(V^j(.,.))_{j \geq T}$ are uniformly locally Lipschitz continuous on $[0, T] \times \Omega$. Let*

$\bar{x} \in W^{1,1}_{\mathrm{loc}}(\mathbb{R}^+_{t_0}; A)$ be such that for any $j \geq T$ the restriction $(\bar{x}, \bar{z})|_{[t_0,j]}(.)$ is a minimizer for problem $\mathcal{M}(\phi^j, j)$ with initial datum (t_0, x_0). Let, for every $j \in \mathbb{N}^+$, p_j, q_j, ν_j, and μ_j be as in the conclusion of Step 2 for the problem $\mathcal{M}(\phi^j, j)$.

Then:

(i) $(p_j)_{j\geq T}$ and $(q_j)_{j\geq T}$ are uniformly bounded on $[t_0, T]$;
(ii) the total variation of the measures $(\tilde{\mu}_j)_{j\geq T}$ on $[t_0, T]$ is uniformly bounded, where $\tilde{\mu}_j$ is defined by $\tilde{\mu}(dt) = \left|\nu_j(t)\right| \mu_j(dt)$.

Proof Since $\bar{x}(.)$ is continuous, hence locally bounded, by the uniform local Lipschitz continuity of $(V^j)_j$ we deduce that

$$\sup \left\{ |\xi| \mid \xi \in \bigcup_{t\in[t_0,T]} \widetilde{\partial}_x V^j(t, \bar{x}(t)) \cup \partial_x^+ V^j(t_0, \bar{x}(t_0)),\ j \geq T \right\} < +\infty. \tag{3.35}$$

By Step 2-(c), (e) we know that

$$-\, q_j(T) \in \partial\phi^j(\bar{x}(T)),\ -q_j(t_0) = -p_j(t_0) \in \partial_x^+ V^j(t_0, \bar{x}(t_0)),$$

and

$$-\, q_j(t) \in \widetilde{\partial}_x V^j(t, \bar{x}(t)) \qquad \text{a.e. } t \in]t_0, T]$$

for all $j \geq T$. Since q_j are right continuous on $]t_0, T[$, from (3.35), it follows that

$$(\|q_j\|_{\infty,[t_0,T]})_{j\geq T} \text{ is bounded.} \tag{3.36}$$

Now, recalling Lemma 1.3.2, from Step 2 and our assumptions, it follows that there exists $\xi \in L^1_{\mathrm{loc}}(\mathbb{R}^+_{t_0}; \mathbb{R}^+_0)$ such that $|p_j'(t)| \leq \xi(t) \left|q_j(t)\right|$ for a.e. $t \in [t_0, T]$ and all $j \geq T$. Hence, in view of (3.36),

$$(\|p_j\|_{\infty,[t_0,T]})_{j\geq T} \text{ is bounded.} \tag{3.37}$$

So, the conclusion *(i)* follows. Also, since $q_j(t) = p_j(t) + \eta_j(t)$, from (3.36) and (3.37) we deduce that

$$(\|\eta_j\|_{\infty,[t_0,T]})_{j\geq T} \text{ is bounded.} \tag{3.38}$$

Now let $\Gamma := \{s \in [t_0, T] \mid \bar{x}(s) \in \mathrm{bdr}\, A\}$. From the relaxed inward pointing condition, it follows that $\mathrm{int}\, T_A^C(\bar{x}(t))$ is nonempty for all $t \in \Gamma$ and so $\mathrm{int}\, T_A^C(\bar{x}(t))$ is nonempty for all $t \in [t_0, T]$. Furthermore, this implies that the set valued map $t \rightsquigarrow T_A^C(\bar{x}(t))$ is lower semicontinuous on $[t_0, T]$. Since

$$\Gamma = \left\{s \in [t_0, T] \mid T_A^C(\bar{x}(s)) \neq \mathbb{R}^n\right\},$$

we can apply Lemma 3.3.5 with $\varepsilon = 1$ to conclude that there exists a continuously differentiable function $f :]t_0 - 1, T + 1[\to \mathbb{R}^n$ such that

$$\sup_{q \in N_A^C(\bar{x}(t)) \cap \mathbb{S}} \langle q, f(t) \rangle \leq -\varepsilon \qquad \forall t \in \Gamma. \tag{3.39}$$

We remark that the function f does not depend on j but only on $\bar{x}(.)$ and T. Then, from (3.39) we deduce that for all $j \geq T$,

$$\begin{aligned}
&\int_{[t_0,T]} \langle f(s), \nu_j(s) \rangle \, d\mu_j(s) \\
&= \int_{[t_0,T] \cap \{s \mid \nu_j(s) \neq 0\}} \langle f(s), \nu_j(s) \rangle \, d\mu_j(s) \\
&= \int_{[t_0,T] \cap \{s \mid \nu_j(s) \neq 0\}} \langle f(s), \frac{\nu_j(s)}{|\nu_j(s)|} \rangle \, |\nu_j(s)| \, d\mu_j(s) \\
&\leq - \int_{[t_0,T] \cap \{s \mid \nu_j(s) \neq 0\}} |\nu_j(s)| \, d\mu_j(s) \\
&= - \int_{[t_0,T]} |\nu_j(s)| \, d\mu_j(s).
\end{aligned}$$

So,

$$\int_{[t_0,T]} |\nu_j(s)| \, d\mu_j(s) \leq \int_{[t_0,T]} \langle -f(s), \nu_j(s) \rangle \, d\mu_j(s). \tag{3.40}$$

Furthermore, from (3.38), integrating by parts, we obtain that, for some constant $C \geq 0$ and all $j \geq T$,

$$\begin{aligned}
&\int_{[t_0,T]} \langle -f(s), \nu_j(s) \rangle \, d\mu_j(s) \\
&= \int_{[t_0,T]} -f(s) \, d\eta_j(s) \\
&= -\eta_j(T) f(T) + \int_{[t_0,T]} \eta_j(s) f'(s) ds \\
&\leq C \left(\|f\|_{\infty,[t_0,T]} + (T - t_0) \left\| f' \right\|_{\infty,[t_0,T]} \right).
\end{aligned} \tag{3.41}$$

Now, since f does not depend on $j \in \mathbb{N}$, from (3.40) and (3.41) we deduce *(ii)*. □

Let k be an integer such that $k > t_0$. Applying Lemma 3.3.6 to problems $\mathscr{M}(\phi^j, j)$, we known that $(p_j)_{j \geq k}$ and $(q_j)_{j \geq k}$ are uniformly bounded on $[t_0, k]$. Furthermore, for some $\xi \in L^1_{\text{loc}}(\mathbb{R}_0^+; \mathbb{R}_0^+)$ and a.e. $t \geq t_0$, we have $|p_j'(t)| \leq \xi(t) \left| q_j(t) \right|$ for all j. So, by the Ascoli-Arzelà and the Dunford-Pettis Theorems we have, taking a subsequence and keeping the same notation, that there exists an absolutely continuous function $p^k : [t_0, k] \to \mathbb{R}^n$ such that $p_j \to p^k$ uniformly on $[t_0, k]$ and $p_j' \rightharpoonup (p^k)'$ in $L^1([t_0, k]; \mathbb{R}^n)$. Furthermore, from Lemma 3.3.6 again, we known that $(\eta_j)_{j\geq k}$ is uniformly bounded on $[t_0, k]$ and the total variation of such functions is uniformly bounded on $[t_0, k]$. So, applying Helly's Selection Theorem, taking a subsequence and keeping the same notation, we deduce that there exists a function of bounded variation η^k on $[t_0, k]$ such that $\eta_j \to \eta^k$ pointwise on $[t_0, k]$ (notice that since $\eta_j(t_0) = 0$ for all j then $\eta^k(t_0) = 0$). Furthermore, from Lemma 3.3.6:*(ii)* we deduce that there exists a nonnegative measure μ^k on $[t_0, k]$ such that, by further extraction of a subsequence, $\tilde{\mu}_j \rightharpoonup^* \mu^k$ in $C([t_0, k]; \mathbb{R})^*$, where $\tilde{\mu}_j(dt) = |\nu_j(t)|\, \mu_j(dt)$. Let

$$\gamma_j(t) := \begin{cases} \dfrac{\nu_j(t)}{|\nu_j(t)|} & \text{if } \nu_j(t) \neq 0, \\ 0 & \text{otherwise.} \end{cases}$$

Since $\gamma_j(t) \in \operatorname{cl}\operatorname{co} N_A(\bar{x}(t)) \cap \mathbb{B}$ for $\tilde{\mu}_j$–a.e. $t \in [t_0, k]$ is a Borel measurable selection, applying Lemma 3.3.4, we deduce that, for a subsequence j_i, there exists a Borel measurable function ν^k such that

$$\nu^k(.) \in \operatorname{cl}\operatorname{co} N_A(\bar{x}(.)) \cap \mathbb{B} \qquad \mu^k - \text{a.e. on } [t_0, k]$$

and for all $\phi \in C([t_0, k]; \mathbb{R}^n)$

$$\int_{[t_0,k]} \langle \phi(s), \gamma_{j_i}(s)\rangle \, d\tilde{\mu}_{j_i}(s) \to \int_{[t_0,k]} \langle \phi(s), \nu^k(s)\rangle \, d\mu^k(s) \qquad \text{as } i \to +\infty. \tag{3.42}$$

Now, since for $t \in]t_0, k]$

$$\begin{aligned} \eta_{j_i}(t) &= \int_{[t_0,t]} \nu_{j_i}(s)\, d\mu_{j_i}(s) = \int_{[t_0,t]\cap\{s \mid \nu_{j_i}(s)\neq 0\}} \nu_{j_i}(s)\, d\mu_{j_i}(s) \\ &= \int_{[t_0,t]} \gamma_{j_i}(s)\, d\tilde{\mu}_{j_i}(s), \end{aligned}$$

from (3.42) it follows that for all $t \in]t_0, k]$

$$\eta^k(t) = \int_{[t_0,t]} \nu^k(s)\, d\mu^k(s).$$

By Mazur's Theorem and standard machinery, using the closedness of $\partial_x^+ V(t_0, x_0)$, $\widetilde{\partial}_x V(t, \bar{x}(t))$ and convexity in (3.29), passing to the limit in (3.32), (3.29), and (3.31) on $[t_0, k]$, and in (3.33) and (3.34) on $]t_0, k]$, we obtain condition *(iv)* on $[t_0, k]$, inclusions *(ii)* on $[t_0, k]$, *(iii)* and *(v)* at t_0 and on $]t_0, k]$.

STEP 4: LIMITING. Consider now the interval $[t_0, k+1]$. By the same argument as in the first step, taking suitable subsequences $(p_{j_{i_l}})_l \subset (p_{j_i})_i$ and $(\eta_{j_{i_l}})_l \subset (\eta_{j_i})_i$, we deduce that there exist an absolutely continuous function p^{k+1}, a function of bounded variation η^{k+1}, and a nonnegative measure μ^{k+1} which satisfy condition *(iv)* on $[t_0, k+1]$, inclusions *(ii)* on $[t_0, k+1]$, *(iii)* and *(v)* at t_0 and on $]t_0, k+1]$. Moreover

$$\begin{cases} p_{j_{i_l}} \to p^{k+1} \text{ uniformly on } [t_0, k+1], \\ p'_{j_{i_l}} \rightharpoonup (p^{k+1})' \text{ in } L^1([t_0, k+1]; \mathbb{R}^n), \\ p^{k+1}|_{[t_0,k]} = p^k, \end{cases}$$

and for all $t \in [t_0, k+1]$

$$\eta_{j_{i_l}}(t) \to \eta^{k+1}(t) = \begin{cases} \int_{[t_0,t]} \nu^{k+1}(s)\, d\mu^{k+1}(s) & t \in]t_0, k+1], \\ 0 & t = t_0, \end{cases}$$

where $\nu^{k+1}(t) \in \operatorname{cl} \operatorname{co} N_A(\bar{x}(t)) \cap \mathbb{B}$ for μ^{k+1} – a.e. $t \in [t_0, k+1]$ is a Borel measurable selection. Furthermore, since $\eta^{k+1}|_{[t_0,k]} = \eta^k$ and $\mu^{k+1}|_{[t_0,k]} = \mu^k$, we have that

$$\nu^{k+1}|_{[t_0,k]} = \nu^k \qquad \mu^k\text{-a.e. on } [t_0, k].$$

We see that the functions p^{k+1}, η^{k+1}, and ν^{k+1} extend the functions p^k, η^k, and ν^k respectively, and measure μ^{k+1} extends measure μ^k.

Repeating the argument above for any interval $[t_0, k+s]$ with $s \in \mathbb{N}$, we can extend p^k, η^k, ν^k and μ^k to the whole interval $\mathbb{R}^+_{t_0}$, extracting every time a subsequence of the previously constructed subsequence. Finally, we conclude that there exists a locally absolutely continuous function $p : \mathbb{R}^+_{t_0} \to \mathbb{R}^n$, a function of locally bounded variation $\eta : \mathbb{R}^+_{t_0} \to \mathbb{R}$, a nonnegative measure μ on $\mathbb{R}^+_{t_0}$, and a Borel measurable selection $\nu(t) \in \operatorname{cl} \operatorname{co} N_A(\bar{x}(t)) \cap \mathbb{B}$ for μ – a.e. $t \in \mathbb{R}^+_{t_0}$ satisfying the conclusion of the theorem. □

Chapter 4
Hamilton-Jacobi-Bellman Equations

Abstract In this chapter, we investigate a class of HJB equations defined over infinite horizon. Our analysis is conducted under a set of mild assumptions, which are sufficient to derive a representation theorem tailored to the case of fiber-convex Hamiltonians. This representation result plays a crucial role in the subsequent development, as it enables us to establish a uniqueness property for weak solutions that vanish at infinity. Particular attention is given to Hamiltonians that are time-measurable, which is a typical features in many control models.

Keywords Hamilton-Jacobi-Bellman equations · Epigraphical representation · Weak solutions · Uniqueness of solutions

4.1 Representation of Fiber-Convex and Time-Measurable Hamiltonians

This section develops a representation result for a broad class of Hamiltonians that are fiber-convex and time-measurable, providing the foundational tools needed to handle such irregular cases within the HJB framework. We denote: for any nonempty subset $E \subset \mathbb{R}^m$, $\|E\| := \sup_{k\in E} |k| \in [0, +\infty]$; $\mathscr{C}^m$ stands for the family of all nonempty closed convex subsets in $\mathbb{R}^m$; we write $J \in \mathscr{C}_b^m$ if J is bounded and $J \in \mathscr{C}^m$. Notice that $d_{\mathscr{H}}(J, K) < +\infty$ for any $J, K \in \mathscr{C}_b^m$.

4.1.1 Parametrization of Set-Valued Maps

Before proceeding to the main representation results, we establish key parametrization properties for set-valued maps that will be essential for our subsequent analysis of Hamiltonian representations. Next, we state the main result of this section on parametrization of convex sets.

V. Basco, *Nonsmooth Constrained Optimal Control on Infinite Horizon*,
SpringerBriefs in Mathematics, https://doi.org/10.1007/978-3-032-08938-0_4

Theorem 4.1.1 *Let I be a closed interval of $\mathbb{R}_0^+$ and $Q : I \times \mathbb{R}^k \rightsquigarrow \mathbb{R}^m$ be a set-valued map such that $Q(t, x) \in \mathscr{C}^m$ for all $(t, x) \in I \times \mathbb{R}^k$, $Q(., x)$ is measurable for all $x \in \mathbb{R}^k$, and:*

$$\forall t \in I,\ \forall r > 0,\ \exists c_r(t) > 0 :\ Q(t, x) \subset Q(t, y) + c_r(t)|x - y|\mathbb{B} \quad \forall x, y \in B(0, r). \tag{4.1}$$

Then, for any set-valued map $\delta : I \times \mathbb{R}^k \rightsquigarrow \mathbb{R}^m$, with nonempty closed values and $\delta(., x)$ measurable for all $x \in \mathbb{R}^k$, there exist two functions $\phi : I \times \mathbb{R}^k \times \mathbb{R}^m \to \mathbb{R}^m$ and $\eta : I \times \mathbb{R}^k \to]0, +\infty[$ satisfying

$$\eta(t, x) = \begin{cases} \|\delta(t, x)\| & \text{if } \|\delta(t, x)\| > 0, \\ 1 & \text{otherwise,} \end{cases} \tag{4.2}$$

and:

(i) $\phi(., x, u)$ *and* $\eta(., x)$ *are measurable for all* $x \in \mathbb{R}^k$, $u \in \mathbb{R}^m$;
(ii) *for any* $t \in I$ *and any* $r > 0$

$$|\phi(t, x, u) - \phi(t, y, v)| \le 5m(c_r(t)|x - y| + |\eta(t, x)u - \eta(t, y)v|)$$
$$\forall x, y \in B(0, r),\ \forall u, v \in \mathbb{R}^m;$$

(iii) $\phi(t, x, \mathbb{B}) \subset Q(t, x)$ *for all* $(t, x) \in I \times \mathbb{R}^k$;
(iv) *if* $\delta(t, x) \neq \{0\}$ *and it is bounded, then* $Q(t, x) \cap \delta(t, x) \subset \phi(t, x, \mathbb{B})$.

In particular, if $\delta(., .) \equiv \mathbb{R}^m$, *then*

$$Q(t, x) = \phi(t, x, \mathbb{R}^m) \quad \forall (t, x) \in I \times \mathbb{R}^k. \tag{4.3}$$

We need first the following results.

Lemma 4.1.2 *Let $(\Theta, \mathfrak{A}, \mu)$ be a complete σ-finite measure space. Let X be a complete separable metric space and $\Phi_n : \Theta \rightsquigarrow X$, $n \in \mathbb{N}$, set-valued maps with closed values. Then the maps $\Theta \ni \theta \rightsquigarrow \mathrm{cl}\,(\bigcup_{n \in \mathbb{N}} \Phi_n(\theta))$ and $\Theta \ni \theta \rightsquigarrow \bigcap_{n \in \mathbb{N}} \Phi_n(\theta)$ are measurable.*

Proof See Sect. A.3.3. □

Lemma 4.1.3 *Let us consider a complete separable metric space $(X, \mathfrak{d}(., .))$, a measurable set-valued map $\Phi : \Theta \rightsquigarrow X$ with closed images, and measurable functions $\phi : \Theta \to X$, $\rho : \Theta \to \mathbb{R}_0^+$. Then the following maps are measurable:*

- *the map* $\theta \rightsquigarrow \{x \in X \mid \mathfrak{d}(x, \phi(\theta)) \le \rho(\theta)\}$;
- *the distance function* $\theta \mapsto \sup_{x \in \Phi(\theta)} \mathfrak{d}(\phi(\theta), x)$.

Proof See Sect. A.3.3. □

Lemma 4.1.4 *Let* $(u, J) \rightsquigarrow P(u, J) := J \cap B(u, 2d_J(u))$ *be the set-valued map defined for any* $u \in \mathbb{R}^k$ *and any* $J \subset \mathbb{R}^k$ *convex nonempty. Then* P *is said to be the projection map and it holds that* $d_{\mathcal{H}}(P(u, J), P(v, K)) \leq 5(d_{\mathcal{H}}(J, K) + |u - v|)$ *for all* $J, K \subset \mathbb{R}^k$ *convex nonempty and all* $u, v \in \mathbb{R}^k$.

Proof See Sect. A.3.3. □

Proof of Theorem 4.1.1 Assume first that $\delta : I \times \mathbb{R}^k \rightsquigarrow \mathbb{R}^m$ is the constant set-valued map $\delta(.,.) \equiv \mathbb{R}^m$. Next, we prove (i)–(iii). Notice that, from our assumptions, Lemma 4.1.3, and since the intersection of measurable set-valued maps is measurable Lemma 4.1.2, we have that

$$\forall (x, u) \in \mathbb{R}^k \times \mathbb{R}^m, \quad t \rightsquigarrow P(u, Q(t, x)) \text{ is measurable,} \tag{4.4}$$

where $P(.,.)$ is the projection map defined in Lemma 4.1.4. Fix $r > 0$. From assumption (4.1) and applying Lemma 4.1.4, for all $t \in I$ and all $x, y \in B(0, r)$

$$\begin{aligned} d_{\mathcal{H}}(P(u, Q(t, x)), P(v, Q(t, y))) &\leq 5(d_{\mathcal{H}}(Q(t, x), Q(t, y)) + |u - v|) \\ &\leq 5(c_r(t)|x - y| + |u - v|). \end{aligned} \tag{4.5}$$

Now, consider the function $\phi : I \times \mathbb{R}^k \times \mathbb{R}^m \to \mathbb{R}^m$ defined by

$$\phi(t, x, u) := S_m \circ P(u, Q(t, x)).$$

Let $t \in I$, $x \in \mathbb{R}^k$, and $u \in \mathbb{R}^m$. By (1.6) immediately follows that $\phi(t, x, u) \in Q(t, x)$. In particular, (iii) holds. Moreover, let $w \in Q(t, x)$. Since $Q(t, x) \cap B(w, 2d(w, Q(t, x))) = \{w\}$, then $\phi(t, x, w) = S_m \circ P(w, Q(t, x)) = w$. So, (4.3) is proved. From the m-Lipschitz continuity of $S_m(.)$ and (4.5), it follows that for all $t \in I$,

$$\begin{aligned} |\phi(t, x, u) - \phi(t, y, v)| &\leq m d_{\mathcal{H}}(P(u, Q(t, x)), P(v, Q(t, y))) \\ &\leq 5m(c_r(t)|x - y| + |u - v|) \end{aligned} \tag{4.6}$$

for all $x, y \in B(0, r)$ and all $u, v \in \mathbb{R}^m$. Hence, recalling (4.4), (4.6), and the continuity of $S_k(.)$, (i) and (ii) follows.

Now, consider a set-valued map $\delta : I \times \mathbb{R}^k \rightsquigarrow \mathbb{R}^m$ with nonempty closed values and $\delta(., x)$ measurable for all $x \in \mathbb{R}^k$. From Lemma 3.2.4 and since $\delta(., x)$ is measurable for any $x \in \mathbb{R}^k$, we have that the map $t \to \|\delta(t, x)\| \in [0, +\infty]$ is measurable for any $x \in \mathbb{R}^k$. Define, for any $x \in \mathbb{R}^k$, the measurable set $\Lambda(x) := \{t \in I \mid \|\delta(t, x)\| \in \{0, +\infty\}\}$ and denote for all $(t, x) \in I \times \mathbb{R}^k$

$$\eta(t, x) := \chi_{\Lambda(x)}(t) + \chi_{\Lambda(x)^c}(t) \, \|\delta(t, x)\| .$$

Then, (4.2) holds and, from our assumptions, the map $t \mapsto \eta(t, x)$ is measurable for all $x \in \mathbb{R}^k$. Consider the map $\phi : I \times \mathbb{R}^k \times \mathbb{R}^m \to \mathbb{R}^m$ defined by

$$\phi(t, x, u) := S_m \circ P(\eta(t, x)u, Q(t, x)).$$

Arguing in the same way as above, the statements (i)-(iii) holds. Next we show (iv). Assume that $\delta(t, x)$ is bounded and $\delta(t, x) \neq \{0\}$. From the definition of $\eta(t, x)$, we have that $\eta(t, x) = \|\delta(t, x)\| > 0$. Now, if $Q(t, x) \cap \delta(t, x) = \emptyset$, then (iv) holds. Otherwise, let $w \in Q(t, x) \cap \delta(t, x)$. Then, there exists $\hat{\delta} \in [0, \|\delta(t, x)\|]$ and $|\hat{w}| = 1$ satisfying $w = \hat{w}\hat{\delta}$. We have $w = (\hat{w}\frac{\hat{\delta}}{\eta(t,x)})\eta(t, x)$. Since $\left|\hat{w}\frac{\hat{\delta}}{\eta(t,x)}\right| \leq 1$, it follows $\phi(t, x, \hat{w}\frac{\hat{\delta}}{\eta(t,x)}) = w$. Thus, (iv) holds. □

Remark 4.1.5 Let $(t, x) \in I \times \mathbb{R}^n$.

(i) From the proof of Theorem 4.1.1, it follows that $d_{\mathscr{H}}(Q(t, x) \cap \delta(t, x), \phi(t, x, \mathbb{B})) \leq 10m \|\delta(t, x)\|$. Indeed, for any $\gamma \in Q(t, x) \cap \delta(t, x)$, any $\theta \in \phi(t, x, \mathbb{B})$, and $u \in \mathbb{B}$, we have:

$$\begin{aligned}
|\gamma - \theta| &= |\gamma - S_m \circ P(\|\delta(t, x)\| u, Q(t, x))| \\
&= |S_m \circ P(\gamma, Q(t, x)) - S_m \circ P(\|\delta(t, x)\| u, Q(t, x))| \\
&\leq 5m|\gamma - \|\delta(t, x)\| u| \\
&\leq 10\, m \, \|\delta(t, x)\| .
\end{aligned}$$

(ii) Using the triangle inequality and the fact that $\phi = S_m(P(\eta(t, x)u, Q(t, x))) \in B(\eta(t, x)u, 2d_{Q(t,x)}(\eta(t, x)u))$, for all $u \in \mathbb{B}$

$$\begin{aligned}
|\phi(t, x, u)| &\leq \eta(t, x)|u| + 2\text{dist}(\eta(t, x)u, Q(t, x)) \\
&\leq 3\eta(t, x) + 2\text{dist}(0, Q(t, x)).
\end{aligned}$$

From Theorem 4.1.1 we get the following corollary.

Corollary 4.1.6 *Assume the assumptions of Theorem 4.1.1 and that for all $x \in \mathbb{R}^k$ there exists $r(., x) : I \to]0, +\infty[$ measurable such that $Q(t, x) \subset r(t, x)\mathbb{B}$ for all $(t, x) \in I \times \mathbb{R}^k$.*

Then there exists a function $\phi : I \times \mathbb{R}^k \times \mathbb{R}^m \to \mathbb{R}^m$ such that

$$Q(t, x) = \phi(t, x, \mathbb{B}) \quad \forall (t, x) \in I \times \mathbb{R}^k,$$

and:

(i) $\phi(., x, u)$ is measurable for all $x \in \mathbb{R}^k$, $u \in \mathbb{R}^m$;
(ii) for any $t \in I$ and any $r > 0$

$$\begin{aligned}
&|\phi(t, x, u) - \phi(t, y, v)| \leq 5m(c_r(t)|x - y| + |r(t, x)u - r(t, y)v|) \\
&\forall x, y \in B(0, r), \ \forall u, v \in \mathbb{R}^m.
\end{aligned}$$

Proof All the conclusions follows from Theorem 4.1.1 by choosing $\delta(., .) = r(., .)\mathbb{B}$. □

4.1.2 Representation of Convex Hamiltonians

Building upon the parametrization results for set-valued maps, we now focus specifically on the representation of convex Hamiltonians. The convexity structure allows for more refined analysis and provides stronger representation results that are particularly useful in applications. This subsection establishes the fundamental representation theorem that will serve as the basis for our subsequent regularity analysis.

Let I be a closed interval of $\mathbb{R}_0^+$ and $H : I \times \mathbb{R}^n \times \mathbb{R}^n \to \mathbb{R}$. In the following, for any $(t, x) \in I \times \mathbb{R}^n$, we denote by $H^*(t, x, .) : \mathbb{R}^n \to [-\infty, +\infty]$ the Fenchel transform of the function $H(t, x, .)$ and we define the set-valued maps $D : I \times \mathbb{R}^n \rightsquigarrow \mathbb{R}^n$, $Ep : I \times \mathbb{R}^n \rightsquigarrow \mathbb{R}^{n+1}$, and $Gr : I \times \mathbb{R}^n \rightsquigarrow \mathbb{R}^{n+1}$ respectively by:

$$\begin{aligned} D(t, x) &:= \text{dom } H^*(t, x, .), \\ Ep(t, x) &:= \text{epi } H^*(t, x, .), \\ Gr(t, x) &:= \text{graph } H^*(t, x, .). \end{aligned}$$

Assumption 4.1.7 Consider the conditions on H:

(I) $t \mapsto H(t, x, p)$ is measurable for any $x, p \in \mathbb{R}^n$ and $p \mapsto H(t, x, p)$ is convex for all $t \in I$ and $x \in \mathbb{R}^n$;

(II) for all $r > 0$ there exists $C_r : I \to \mathbb{R}_0^+$ measurable such that

$$|H(t, x, p) - H(t, y, p)| \leq C_r(t)(1 + |p|)|x - y|$$

for all $t \in I$, $x, y \in B(0, r)$, and $p \in \mathbb{R}^n$;

(III) there exists $\tilde{c} : I \to \mathbb{R}_0^+$ measurable such that

$$|H(t, x, p) - H(t, x, q)| \leq \tilde{c}(t)(1 + |x|)|p - q|$$

for all $t \in I$ and $x, p, q \in \mathbb{R}^n$;

(IV) there exists $\bar{\gamma} : I \times \mathbb{R}^n \to \mathbb{R}_0^+$ such that $\bar{\gamma}(., x)$ is measurable for any $x \in \mathbb{R}^n$ and for all $(t, x) \in I \times \mathbb{R}^n$ and all $\bar{q} \in \text{cl } D(t, x)$, there exists $\varepsilon > 0$ such that

$$\sup_{q \in D(t,x) \cap B(\bar{q}, \varepsilon)} |H^*(t, x, q)| \leq \bar{\gamma}(t, x).$$

Theorem 4.1.8 (Representation) *Consider Assumption 4.1.7:(I), (II), (IV).*

Then there exists a function $\phi : I \times \mathbb{R}^n \times \mathbb{R}^{n+1} \to \mathbb{R}^{n+1}$,

$$\phi(t, x, u) =: (f(t, x, u), \ell(t, x, u)) \in \mathbb{R}^n \times \mathbb{R},$$

satisfying:

(i) $f(., x, u)$ and $\ell(., x, u)$ are measurable for all $x \in \mathbb{R}^n$ and $u \in \mathbb{R}^{n+1}$;

(ii) for all $t \in I$ and $x, p \in \mathbb{R}^n$,

$$H(t,x,p) = \sup_{u \in \mathbb{R}^{n+1}} \{\langle p, f(t,x,u)\rangle - \ell(t,x,u)\};$$

(iii) for all $t \in I$ and $r > 0$,

$$|f(t,x,u) - f(t,y,v)| \leq 10(n+1)(C_r(t)|x-y| + |u-v|),$$
$$|\ell(t,x,u) - \ell(t,y,v)| \leq 10(n+1)(C_r(t)|x-y| + |u-v|),$$

for every $x, y \in B(0,r)$ and $u, v \in \mathbb{R}^{n+1}$.

If, in addition, condition Assumption 4.1.7:(III) holds, then the statements (ii)-(iii) are replaced, respectively, by:

(ii)' for all $t \in I$ and $x, p \in \mathbb{R}^n$,

$$H(t,x,p) = \sup_{u \in \mathbb{B} \subset \mathbb{R}^{n+1}} \{\langle p, f(t,x,u)\rangle - \ell(t,x,u)\};$$

(iii)' for all $t \in I$ and $r > 0$,

$$|f(t,x,u) - f(t,y,v)| \leq 10(n+1)(C_r(t)|x-y| + |\eta(t,x)u - \eta(t,y)v|),$$
$$|\ell(t,x,u) - \ell(t,y,v)| \leq 10(n+1)(C_r(t)|x-y| + |\eta(t,x)u - \eta(t,y)v|),$$

for all $x, y \in B(0,r)$ and $u, v \in \mathbb{R}^{n+1}$, where

$$\eta(t,.) := \tilde{c}(t)(1 + |\,.\,|) + \bar{\gamma}(t,.);$$

and the additional conditions also hold:

(iv)' $D(t,x) = f(t,x,\mathbb{B})$ for all $t \in I$ and $x \in \mathbb{R}^n$;
(v)' $Gr(t,x) \subset \phi(t,x,\mathbb{B})$ for all $t \in I$ and $x \in \mathbb{R}^n$.

Moreover, if Assumption 4.1.7:(IV) is replaced by the weaker

$$\sup_{q \in D(t,x) \cap B(\bar{q},\varepsilon)} H^*(t,x,q) \leq \bar{\gamma}(t,x),$$

then the above statements holds true with

$$\eta(t,.) := (\tilde{c}(t) + C_r(t))(1 + |\,.\,|) + |H(t,0,0)| + \bar{\gamma}(t,.).$$

Before to give a proof of Theorem 4.1.8, we show some intermediate results.

Lemma 4.1.9 *Consider Assumption 4.1.7:(I), (II) and let $(t,x) \in I \times \mathbb{R}^n$. Then:*

(i) $Ep(t,x)$ *and* $D(t,x)$ *are nonempty convex, and* $Ep(t,x)$ *is closed.*

Moreover, if, in addition, the Assumption 4.1.7:(III) holds, then:

(ii) $D(t,x) \subset \tilde{c}(t)(1+|x|)\mathbb{B}$;
(iii) $H^*(t,.,.)$ *is lower semicontinuous.*

Proof Let $(t,x) \in I \times \mathbb{R}^n$ and $p \in \mathbb{R}^n$. Since $\partial_p H(t,x,p) \neq \emptyset$, there exists $\xi \in \partial_p H(t,x,p)$. From Lemma 1.2.3 it follows that $H^*(t,x,\xi) \neq +\infty$. Then $D(t,x)$ is nonempty and in particular $Ep(t,x)$ as well. Moreover, they are convex as consequence of our assumptions and Lemma 1.2.3. $Ep(t,x)$ is closed since the Fenchel transform is lower semicontinuous. So, (i) is proved.

Now, assume Assumption 4.1.7:(III) and let $\xi \in D(t,x)$. If there exists $p \in \partial_v H^*(t,x,\xi)$, then, by Lemma 1.2.3, $\xi \in \partial_p H(t,x,p)$. By Assumption 4.1.7:(III), $H(t,x,.)$ is $c(t)(1+|x|)$-Lipschitz. Thus $|\xi| \leq \tilde{c}(t)(1+|x|)$. If $\partial_v H^*(t,x,\xi) = \emptyset$, then, using the separation theorem, we deduce that there exists

$$0 \neq (p,0) \in N_{Ep(t,x)}\left(\xi, H^*(t,x,\xi)\right).$$

By Lemma 1.2.2:*(ii)*, we can find a sequence $v_i \to \xi$ such that $(v_i, H^*(t,x,v_i)) \to (\xi, H^*(t,x,\xi))$, when $i \to +\infty$ and $\partial_v H^*(t,x,v_i) \neq \emptyset$. By the first part of the proof $|v_i| \leq \tilde{c}(t)(1+|x|)$. Consequently, $|\xi| \leq \tilde{c}(t)(1+|x|)$, and *(ii)* holds.

Next, we show *(iii)*. Let $t \in I$ and $(x,v) \in \mathbb{R}^n \times \mathbb{R}^n$. Consider a sequence $(x_i, v_i) \to (x,v)$ such that

$$\liminf_{(y,w)\to(x,v)} H^*(t,y,w) = \lim_{i\to+\infty} H^*(t,x_i,v_i).$$

Let $\varepsilon > 0$. If this lower limit is finite, then for some $M \geq 0$ and all large $i \in \mathbb{N}$, $H^*(t,x_i,v_i) \leq M$. Fix any $p \in \mathbb{R}^n$. So,

$$\langle v,p\rangle - H(t,x,p) \leq \langle v_i,p\rangle - H(t,x_i,p) + \varepsilon \leq M + \varepsilon \text{ for all large } i \in \mathbb{N}.$$

Since p is arbitrary, $H^*(t,x,v) \leq M+\varepsilon$. Consider $p \in \mathbb{R}^n$ such that $H^*(t,x,v) \leq \langle v,p\rangle - H(t,x,p) + \varepsilon$. Since $H(t,.,.)$ is continuous as consequence of our assumptions, for all large $i \in \mathbb{N}$,

$$\langle v,p\rangle - H(t,x,p) \leq \langle v_i,p\rangle - H(t,x_i,p) + \varepsilon \leq H^*(t,x_i,v_i) + \varepsilon.$$

Hence $H^*(t,x,v) \leq \lim_{i\to+\infty} H^*(t,x_i,v_i) + \varepsilon$. The conclusion follows from the arbitrariness of ε and (x,v). □

Lemma 4.1.10 *Consider Assumption 4.1.7:(I), (II), (IV).*
Then:

(i) $D(t,x) \in \mathscr{C}^n$;

(ii) for all $t \in I$ and $r > 0$,

$$D(t,x) \subset D(t,y) + C_r(t)|x-y|\mathbb{B} \quad \forall x, y \in B(0,r).$$

Proof Let $(t,x) \in I \times \mathbb{R}^n$. Next we show that $D(t,x)$ is closed. Consider a sequence $q_i \in D(t,x)$ converging to $\tilde{q} \in \mathbb{R}^n$. Since $\tilde{q} \in \text{cl}\, D(t,x)$, from Assumption 4.1.7:(IV), there exist $\varepsilon > 0$ and $M > 0$ such that $|H^*(t,x,q_i)| \le M$ for all $q_i \in B(\tilde{q},\varepsilon)$. Hence, since the Fenchel transform $q \mapsto H^*(t,x,q)$ is lower semicontinuous, $M \ge \liminf_{i\to+\infty} H^*(t,x,q_i) \ge H^*(t,x,\tilde{q})$. So, $\tilde{q} \in D(t,x)$, and recalling Lemma 4.1.9, the assertion (i) is proved.

Now, to show (ii), suppose by contradiction that there exist $t \in I$, $r > 0$, $x, y \in B(0,r)$, $w \in D(t,x)$, and $\eta > C_r(t)$ such that

$$D(t,y) \cap B(w, \eta|x-y|) = \emptyset.$$

We divide the proof into three steps.

STEP 1: Applying Lemma 4.1.9 and (i), the set $D(t,x)$ is closed and convex. Let $\bar{q} \in D(t,y)$ be the projection of w onto $D(t,y)$ and put $z := (w-\bar{q})/|w-q|$. We have that z is a proximal normal to $D(t,y)$ at $\bar{q}$, i.e., there exists $\bar{\lambda} > \eta|x-y|$ such that $d_{D(t,y)}(\bar{q}+\bar{\lambda}z) = \bar{\lambda}$. Consider the hyperplane $\{\xi \in \mathbb{R}^n \mid \langle z,\xi\rangle = \langle z,\bar{q}\rangle\}$. Since $D(t,y)$ is convex and z is a proximal normal, we have that $D(t,y) \subset \{\xi \in \mathbb{R}^n \mid \langle z,\xi\rangle \le \langle z,\bar{q}\rangle\}$. Moreover, from (1.1), $B(w,\eta|x-y|) \subset \text{int}\, B(\bar{q}+\bar{\lambda}z,\bar{\lambda}) \subset D(t,y)^c$, and we get

$$\langle z,q\rangle \le \langle z,\bar{q}\rangle < \langle z, w+\eta|x-y|h\rangle \quad \forall q \in D(t,y),\ \forall h \in \mathbb{B}. \tag{4.7}$$

Notice that, applying Lemma 1.1.1:*(vi)*, $z \in N_{D(t,y)}(\bar{q})$. Hence, using (1.2), we have $(z,0) \in N_{\text{epi}\, H^*(t,y,.)}(\bar{q}, H^*(t,y,\bar{q}))$.

STEP 2: From Step 1 and applying Lemma 1.2.2:*(i)*, consider two sequences $w_i \in \text{dom}\, H^*(t,y,.)$ and $(p_i,q_i) \in N_{\text{epi}\, H^*(t,y,.)}(w_i, H^*(t,y,w_i))$, with $q_i < 0$, satisfying

$$(p_i,q_i) \to (z,0), \quad (w_i, H^*(t,y,w_i)) \to (\bar{q}, H^*(t,y,\bar{q})). \tag{4.8}$$

So, $(p_i/|q_i|,-1) \in N_{\text{epi}\, H^*(t,y,.)}(w_i, H^*(t,y,w_i))$ for all $i \in \mathbb{N}$. We conclude that $p_i/|q_i| \in \partial H^*(t,y,.)(w_i)$ for all $i \in \mathbb{N}$. Thus, from Lemma 1.2.3,

$$H(t,y,p_i/|q_i|) + H^*(t,y,w_i) = \langle w_i, p_i/|q_i|\rangle \quad \forall i \in \mathbb{N}. \tag{4.9}$$

STEP 3: Using (4.8) and (4.7) with $h = -z/|z|$, we can assume that $\langle w_i,p_i\rangle < \langle w,p_i\rangle - \eta|x-y|$ for all large $i \in \mathbb{N}$. Hence, from Assumption 4.1.7:(II) and recalling that $w \in D(t,x)$, we get for all large $i \in \mathbb{N}$

$$\begin{aligned}
&\langle w_i, p_i\rangle - |q_i|\, H(t, y, p_i/|q_i|) \\
&< \langle w, p_i\rangle - \eta|x-y| - |q_i|H(t, x, p_i/|q_i|) + (|q_i| + |p_i|)C_R(t)|x-y| \\
&= |q_i|(\langle w, p_i/|q_i|\rangle - H(t, x, p_i/|q_i|)) + ((|q_i| + |p_i|)C_r(t) - \eta)|x-y| \\
&\leq |q_i|H^*(t, x, w) + ((|q_i| + |p_i|)C_r(t) - \eta)|x-y|.
\end{aligned}$$

So, by (4.9), for all large $i \in \mathbb{N}$

$$|q_i|H^*(t, y, w_i) < |q_i|H^*(t, x, w) + ((|q_i| + |p_i|)C_r(t) - \eta)|x-y|. \tag{4.10}$$

From Assumption 4.1.7:(IV), the lower semicontinuity of $H^*(t, y, .)$, and since $w_i \to \bar{q}$, the sequence $(H^*(t, y, w_i))_{i\in\mathbb{N}}$ is bounded. Then, using again (4.8), and passing to the lower limit as $i \to +\infty$ in (4.10) we get $0 \leq (|z|C_r - \eta)|x-y|$. Since $|z| = 1$, $0 \leq C_r(t) - \eta$, and a contradiction follows. □

Lemma 4.1.11 *Assume Assumption 4.1.7:(I), (II), (IV).*
Then:

(i) $t \rightsquigarrow Ep(t, x)$ *is measurable for all* $x \in \mathbb{R}^n$;
(ii) $Ep(t, x) \in \mathscr{C}^{n+1}$ *for all* $t \in I$ *and* $x \in \mathbb{R}^n$;
(iii) for all $t \in I$ *and* $r > 0$,

$$Ep(t, x) \subset Ep(t, y) + 2C_r(t)|x-y|\mathbb{B} \quad \forall x, y \in B(0, r).$$

Proof We notice that, from Assumption 4.1.7:(I), (II), the lower semicontinuity and the convexity of Fenchel transform (cfr. Lemma 1.2.3), it follows immediately that, for any $t \in I$ and $x \in \mathbb{R}^n$, the set-valued map $s \rightsquigarrow Ep(s, x)$ is measurable and $Ep(t, x)$ is nonempty, closed, and convex. So, (i) and (ii) holds.

Now, we show (iii). Fix $x, y \in \mathbb{R}^n$, $t \in I$, and consider $(q, \lambda) \in Ep(t, x)$. Without loss of generality we may assume that $C_r(t)|x-y| \neq 0$. We claim the following: there exists $w \in D(t, y)$ satisfying $(w, \lambda + C_r(t)|x-y|) \in Ep(t, y)$. Indeed, from Assumption 4.1.7:(II), we have that $H(t, x, p) \leq H(t, y, p) + C_r(t)(1 + |p|)|x-y|$ for all $p \in \mathbb{R}^n$, and, from the definition of Fenchel transform,

$$(H(t, y, .) + C_r(t)(1 + |\,.\,|)|x-y|)^*\,(\tilde{q}) \leq H^*(t, x, \tilde{q}) \quad \forall \tilde{q} \in \mathbb{R}^n. \tag{4.11}$$

Now, define $h(.) := -C_r(t)|x-y|$ on $B(0, C_r(t)|x-y|)$ and $+\infty$ elsewhere. Notice that $h(.)$ is a proper lower semicontinuous convex function and $h^*(.) = C_r(t)(1 + |\,.\,|)|x-y|$. Notice that, since $\operatorname{dom} H(t, y, .) = \mathbb{R}^n$ and from Assumption 4.1.7:(I), applying Lemma 1.2.3 we get $(H^*(t, y, .))^* = H(t, y, .)$. So, for all $z \in \mathbb{R}^n$

$$
\begin{aligned}
&\left(\inf_{q_1\in\mathbb{R}^k} H^*(t,y,q_1)+h(.-q_1)\right)^*(z)\\
&:= \sup_{q_2\in\mathbb{R}^k}\left\{\langle q_2,z\rangle - \inf_{q_1\in\mathbb{R}^k}\left\{H^*(t,y,q_1)+h(q_2-q_1)\right\}\right\}\\
&= \sup_{q_2\in\mathbb{R}^k}\left\{\langle q_2,z\rangle + \sup_{q_1\in\mathbb{R}^k}\left\{-H^*(t,y,q_1)-h(q_2-q_1)\right\}\right\}\\
&= \sup_{q_2,q_1\in\mathbb{R}^k}\left\{\langle q_2,z\rangle - H^*(t,y,q_1)-h(q_2-q_1)\right\}\\
&= \sup_{q_1\in\mathbb{R}^k}\left\{\langle q_1,z\rangle - H^*(t,y,q_1)+\sup_{q_2\in\mathbb{R}^k}\left\{\langle q_2-q_1,z\rangle-h(q_2-q_1)\right\}\right\}\\
&= H(t,y,z)+C_r(t)(1+|z|)|x-y|.
\end{aligned}
$$

Since $\operatorname{dom} h = B(0, C_r(t)|x-y|)$ and the function

$$
z \mapsto \left(\inf_{q_1\in\mathbb{R}^k} H^*(t,y,q_1)+h(z-q_1)\right)
$$

is proper lower semicontinuous and convex, passing to the Fenchel transform and using Lemma 1.2.3 we deduce

$$
\begin{aligned}
&\inf_{q_1\in\mathbb{R}^k} H^*(t,y,q_1)+h(q-q_1)\\
&= \inf_{q_1\in B(q,C_r(t)|x-y|)} H^*(t,y,q_1)-C_r(t)|x-y|\\
&= (H(t,y,.)+C_r(t)(1+|\,.\,|)|x-y|)^*(q).
\end{aligned}
$$

Thus, from (4.11), there exists $w\in B(q, C_r(t)|x-y|)$ satisfying

$$
H^*(t,y,w)-C_r(t)|x-y| \le H^*(t,x,q).
$$

Hence, the claim follows. Now, applying Lemma 4.1.10:(ii), $|q-w|\le C_r(t)|x-y|$ because $q\in D(t,x)$ and $w\in D(t,y)$. Finally, since

$$
(q,\lambda) = (w, \lambda + C_r(t)|x-y|) + C_r(t)|x-y|\left(\frac{q-w}{C_r(t)|x-y|}, -1\right),
$$

the statement (iii) follows by the arbitrariness of $(q,\lambda)\in Ep(t,x)$. □

Proposition 4.1.12 *Consider Assumption 4.1.7:(I), (II), (IV).*
Then there exists a function $\phi : I\times\mathbb{R}^n\times\mathbb{R}^{n+1}\to\mathbb{R}^{n+1}$,

$$
\phi(t,x,u) =: (f(t,x,u), \ell(t,x,u)) \in \mathbb{R}^n\times\mathbb{R},
$$

satisfying

$$Ep(t,x) = \phi(t,x,\mathbb{R}^{n+1}) \quad \forall (t,x) \in I \times \mathbb{R}^n, \tag{4.12}$$

and:

(i) *$\phi(.,x,u)$ is measurable for all $x \in \mathbb{R}^n$ and $u \in \mathbb{R}^{n+1}$;*
(ii) *for all $t \in I$ and $r > 0$*

$$|\phi(t,x,u) - \phi(t,y,v)| \leq 10(n+1)(C_r(t)|x-y| + |u-v|)$$
$$\forall x, y \in B(0,r), \ \forall u, v \in \mathbb{R}^{n+1}.$$

If, in addition, Assumption 4.1.7:(III) holds, then

$$Gr(t,x) \subset \phi(t,x,\mathbb{B}) \quad \forall (t,x) \in I \times \mathbb{R}^n,$$

the statement (ii) is replaced by the following:

(ii)' for all $t \in I$ and $r > 0$

$$|\phi(t,x,u) - \phi(t,y,v)| \leq 10(n+1)(C_r(t)|x-y| + |\eta(t,x)u - \eta(t,y)v|)$$
$$\forall x, y \in B(0,r), \ \forall u, v \in \mathbb{R}^{n+1},$$

where $\eta(t,.) := \tilde{c}(t)(1+|\,.\,|) + \bar{\gamma}(t,.)$;

and moreover:

(iii)' $D(t,x) = f(t,x,\mathbb{B})$ for all $t \in I$ and all $x \in \mathbb{R}^n$.

If Assumption 4.1.7:(IV) is replaced by the weaker

$$\sup_{q \in D(t,x) \cap B(\bar{q},\varepsilon)} H^*(t,x,q) \leq \bar{\gamma}(t,x)$$

then the above statements holds true with

$$\eta(t,.) := (\tilde{c}(t) + C_r(t))(1+|\,.\,|) + |H(t,0,0)| + \bar{\gamma}(t,.).$$

Proof The first two statements follows from Theorem 4.1.1 and Lemma 4.1.11, with $Q(t,x) = Ep(t,x)$ and $\delta(t,x) = \mathbb{R}^{n+1}$.

Now, consider Assumption 4.1.7:(III) and let $t \in I$. We get $|(v, H^*(t,x,v))| \leq \tilde{c}(t)(1+|x|) + \bar{\gamma}(t,x)$ for all $x \in \mathbb{R}^n$. So

$$\|Gr(t,x)\| \leq \tilde{c}(t)(1+|x|) + \bar{\gamma}(t,x) \tag{4.13}$$

for all $x \in \mathbb{R}^n$. Hence, the statements (ii)'–(iii)' follows from Theorem 4.1.1 and Lemma 4.1.11, with $Q(t,x) = Ep(t,x)$ and $\delta(t,x) = (\tilde{c}(t)(1+|x|) + \bar{\gamma}(t,x))\mathbb{B}$.

The last statement follows as well since for all $r > 0$, all $x \in B(0,r)$, and all $v \in D(t,x)$, and from the compactness of $D(t,x)$,

$$-H(t,x,0) \le H^*(t,x,v) \le \bar{\gamma}(t,x),$$

and so

$$\begin{aligned}|(v, H^*(t,x,v))| &\le \tilde{c}(t)(1+|x|) + \gamma(t,x) + |H(t,x,0)| \\ &\le \tilde{c}(t)(1+|x|) + C_r(t)(1+|x|) + |H(t,0,0)| + \gamma(t,x)\end{aligned}$$

for all $x \in B(0,r)$. □

Next, we give a proof of Theorem 4.1.8.

Proof of Theorem 4.1.8 The statements (i) and (iii) follows from Proposition 4.1.12.

Next we show (ii). Fix $t \in I$, $x \in \mathbb{R}^n$, and $p \in \mathbb{R}^n$. Consider the function $\phi = (f, \ell)$ from Proposition 4.1.12. Recalling that $(H^*(t,x,.))^* = H(t,x,.)$, from (4.12) it follows that for any $u \in \mathbb{R}^{n+1}$ the pair $(f(t,x,u), \ell(t,x,u))$ lays in $Ep(t,x)$, i.e.,

$$H^*(t,x,f(t,x,u)) \le \ell(t,x,u). \tag{4.14}$$

So, for any $u \in \mathbb{R}^{n+1}$

$$\begin{aligned}&\langle p, f(t,x,u)\rangle - \ell(t,x,u) \\ &\le \langle p, f(t,x,u)\rangle - H^*(t,x,f(t,x,u)) \\ &\le \sup_{v \in \mathbb{R}^{n+1}} \{\langle p, v\rangle - H^*(t,x,v)\} = H(t,x,p).\end{aligned}$$

Then, by arbitrariness of $u \in \mathbb{R}^{n+1}$, we get

$$\sup_{u \in \mathbb{R}^{n+1}} \{\langle p, f(t,x,u)\rangle - \ell(t,x,u)\} \le H(t,x,p).$$

On the other hand, let $v \in D(t,x)$. Since $(v, H^*(t,x,v)) \in Ep(t,x)$, from (4.12), there exists $w \in \mathbb{R}^{n+1}$ such that $(v, H^*(t,x,v)) = (f(t,x,w), \ell(t,x,w))$. So, $\langle p, v\rangle - H^*(t,x,v) = \langle p, f(t,x,w)\rangle - l(t,x,w) \le \sup_{u \in \mathbb{R}^{n+1}}\{\langle p, f(t,x,u)\rangle - \ell(t,x,u)\}$. Hence,

$$\begin{aligned}H(t,x,p) &= \sup_{v \in D(t,x)} \{\langle p, v\rangle - H^*(t,x,v)\} \\ &\le \sup_{u \in \mathbb{R}^{n+1}} \{\langle p, f(t,x,u)\rangle - \ell(t,x,u)\}.\end{aligned}$$

The last statements, assuming that Assumption 4.1.7:(III) holds, can be obtained with the same arguments as above using Proposition 4.1.12. □

Remark 4.1.13 Let $t \in I$ and $x \in \mathbb{R}^n$.

(i) The condition in Assumption 4.1.7:(IV) is equivalent to require that $H^*(t,x,.)$ is locally bounded on its domain. If $H(t,x,.)$ is globally Lipschitz, then one can show that Assumption 4.1.7:(IV) is equivalent to assume that $H^*(t, x, .)$ is bounded on its domain.

(ii) We would like to underline that, if $H(t, x, .)$ is convex, then, from Lemma 4.1.9, the domain $D(t, x)$ of the Fenchel transform $H^*(t, x, .)$ turn out to be bounded under the global Lipschitz Assumption 4.1.7:(III). In particular, we notice that the Lipschitz condition imply the sublinear growth of $H(t, x, .)$. On the other hand, it possible to show that, when $H(t, x, .)$ is merely locally Lipschitz continuous, $D(t, x) = \mathbb{R}^n$ if and only if $H(t, x, .)$ is coercive, i.e., $\lim_{|p|\to+\infty} \frac{H(t,x,p)}{|p|} = +\infty$.

(iii) From Remark 4.1.5:(ii) and (4.13) we get for all $u \in \mathbb{B} \subset \mathbb{R}^{n+1}$:

$$\begin{aligned} |(f(t, x, u), \ell(t, x, u))| &\leq 3\eta(t, x) + 2\text{dist}(0, Ep(t, x)) \\ &\leq 3\eta(t, x) + 2\,\|Gr(t, x)\|, \end{aligned} \tag{4.15}$$

hence $|(f(t, x, u), \ell(t, x, u))| \leq 5(\tilde{c}(t)(1 + |x|) + \bar{\gamma}(t, x))$.

4.2 Value Function Properties

The regularity properties of the value function are intimately connected to the geometric properties of its epigraph. Understanding these connections is crucial for establishing the well-posedness of the HJB equation and for developing numerical approximation schemes. In this section, we investigate the regularity of the epigraph under various assumptions on the problem data, providing conditions that ensure desirable smoothness properties.

In the following we consider a closed nonempty subset $\Omega \subset \mathbb{R}^n$.

Assumption 4.2.1 Consider the Assumption 4.1.7 with the further conditions:

(I) $t \mapsto H(t, x, p)$ is measurable for any $x, p \in \mathbb{R}^n$ and $p \mapsto H(t, x, p)$ is convex for all $t \in I$ and $x \in \mathbb{R}^n$;

(II) there exists $C \in \mathcal{L}_{\text{loc}}$ such that

$$|H(t, x, p) - H(t, y, p)| \leq C(t)(1 + |p|)|x - y|$$

for all $t \in I$, $x, y \in \mathbb{R}^n$, and $p \in \mathbb{R}^n$;

(III) there exists $\tilde{c} \in \mathcal{L}_{\text{loc}}$ such that

$$|H(t, x, p) - H(t, x, q)| \leq \tilde{c}(t)(1 + |x|)|p - q|$$

for all $t \in I$ and $x, p, q \in \mathbb{R}^n$;

(IV) there exist $\varphi, \tilde{\varphi} \in \mathcal{L}_{\text{loc}}$ such that for a.e. $t \geq 0$, for all $x \in \mathbb{R}^n$, and all $q \in D(t, x)$

$$-\varphi(t)(1 + |x|) \leq H^*(t, x, q) \leq \tilde{\varphi}(t)(1 + |x|).$$

For all $t \in \mathbb{R}_0^+$, $x \in \mathbb{R}^n$, and $u \in \mathbb{R}^{n+1}$ we denote by $(f(t,x,u), \ell(t,x,u)) := \phi(t,x,u)$ the representation of the Hamiltonian given by Theorem 4.1.8, and by $\mathcal{U}_\Omega(t,x)$ the (possibly empty) set of all trajectory-control pairs $(\xi, u) : \mathbb{R}_t^+ \to \mathbb{R}^n \times \mathbb{R}^{n+1}$ such that $u(.)$ is measurable and

$$\begin{cases} \xi'(s) = f(s, \xi(s), u(s)),\ u(s) \in \mathbb{B} \subset \mathbb{R}^{n+1} & \text{a.e. } s \in \mathbb{R}_t^+, \\ \xi(t) = x, & \\ \xi(s) \in \Omega & \forall s \in \mathbb{R}_t^+. \end{cases} \tag{4.16}$$

The value function $v : \mathbb{R}_0^+ \times \mathbb{R}^n \to [-\infty, +\infty]$ associated to the representation (f, ℓ) is defined by

$$v(t,x) := \inf \left\{ \int_t^{+\infty} \ell(s, \xi(s), u(s))\, ds \mid (\xi(.), u(.)) \in \mathcal{U}_\Omega(t,x) \right\},$$

where $v(t,x) = +\infty$ if $\mathcal{U}_\Omega(t,x) = \emptyset$, by convention. In the following we consider the outward pointing condition:

Assumption 4.2.2 (*Outward Pointing Condition*) there exist $\eta > 0$, $r > 0$, and $M \geq 0$ such that: for a.e. $t \in \mathbb{R}_0^+$, all $y \in \text{bdr}\, \Omega + \eta\mathbb{B}$, and all

$$v \in D(t,y) \cap \{p \in \mathbb{R}^n \mid \exists n \in N_\Omega(y;\eta),\ \langle p, n\rangle \leq 0\},$$

there exists $w \in D(t,y) \cap B(v, M)$ such that

$$w,\ w - p \in \{p \in \mathbb{R}^n \mid \langle p, n\rangle \geq r,\ \forall n \in N_\Omega(y;\eta)\},$$

where $N_\Omega(x;\eta)$ is as in (2.38).

For all $t \in \mathbb{R}_0^+$, $x \in \mathbb{R}^n$, and $u \in \mathbb{R}^{n+1}$ we put

$$L(t,x,u) := H^*(t,x,f(t,x,u)).$$

Lemma 4.2.3 *Consider Assumption 4.2.1. Then:*

(i) for all $x \in \mathbb{R}^n$ the mappings $f(.,x,.)$ and $\ell(.,x,.)$ are Lebesgue-Borel measurable;

(ii) $|f(t,x,u)| + |\ell(t,x,u)| \leq c(t)(1+|x|)$ for a.e. $t \geq 0$ and for all $x \in \mathbb{R}^n$, $u \in \mathbb{B}$ where $c(t) := 5(\tilde{c}(t) + \varphi(t) + \tilde{\varphi}(t))$;

(iii) for a.e. $t \in \mathbb{R}_0^+$, the set-valued map $\mathbb{R}^n \ni x \rightsquigarrow \{(f(t,x,u), \ell(t,x,u)) \mid u \in \mathbb{B}\}$ is continuous with closed images;

(iv) $|f(t,x,u) - f(t,y,u)| + |\ell(t,x,u) - \ell(t,y,u)| \leq k(t)|x-y|$ for a.e. $t \geq 0$ and for all $x, y \in \mathbb{R}^n$, $u \in \mathbb{B}$ where $k(t) := 10(n+1)C(t)$;

(v) $\{(f(t,x,u), \ell(t,x,u) + r) | u \in \mathbb{B}, r \geq 0\}$ is convex for any $t \in \mathbb{R}_0^+$, $x \in \mathbb{R}^n$.

Proof All the statements *(i)–(iv)* follows from our assumptions, Theorem 4.1.8, Remarks 4.1.5:*(i)*, and 4.1.13:*(iii)*. Notice also that

$$\begin{aligned}&\{(f(t,x,u),\ell(t,x,u)+r)|u\in\mathbb{B},r\geq 0\}\\&=\{(f(t,x,u),\ell(t,x,u))|u\in\mathbb{B}\}+\{(0,r)|r\geq 0\}\\&\subset Ep(t,x)+\{(0,r)|r\geq 0\}=Ep(t,x).\end{aligned}$$

Hence, recalling Theorem 4.1.8 (v)', the last statement follows from the convexity of the epigraph of $H^*(t,x,.)$. □

Next, we recall the definition of locally absolutely continuity for set-valued maps.

Proposition 4.2.4 *Consider Assumptions 4.2.1 and 4.2.2. If*

$$\int_0^{+\infty}\varphi(t)e^{\int_0^t c(s)ds}dt<+\infty,$$

then:

(i) $v(\bar{t},\bar{x})\neq-\infty$ *for any* $(\bar{t},\bar{x})\in\mathbb{R}_0^+\times\Omega$ *and for any* $(\bar{t},\bar{x})\in\mathrm{dom}\ v$ *there exists a feasible trajectory-control pair* (x,u) *starting from* $(\bar{t},\bar{x})$ *such that*

$$\text{the limit }\lim_{T\to+\infty}\int_{\bar{t}}^{T}\ell(t,x(t),u(t))dt\text{ exists}$$

and

$$v(\bar{t},\bar{x})=\int_{\bar{t}}^{+\infty}\ell(t,x(t),u(t))dt:=\lim_{T\to+\infty}\int_{\bar{t}}^{T}\ell(t,x(t),u(t))dt;$$

(ii) v *is lower semicontinuous and for any* $x\in\Omega$

$$\liminf_{\substack{s\to0+\\ y\to_\Omega x}}v(s,y)=v(0,x);\tag{4.17}$$

(iii) there exists a set $C\subset[0,+\infty[$, *with* $\mu_{\mathscr{L}}(C)=0$, *such that for any* $(t,x)\in\mathrm{dom}\,v\cap(([0,+\infty[\backslash C)\times\Omega)$

$$\exists\bar{u}\in\mathbb{B}\subset\mathbb{R}^{n+1},\quad D_{\uparrow}v(t,x)(1,f(t,x,\bar{u}))\leq-\ell(t,x,\bar{u});\tag{4.18}$$

(iv) there exists a set $C\subset]0,+\infty[$, *with* $\mu_{\mathscr{L}}(]0,+\infty[\backslash C))=0$, *such that for any* $(t,x)\in\mathrm{dom}\,v\cap(C\times\mathrm{int}\,\Omega)$

$$\forall u\in\mathbb{B}\subset\mathbb{R}^{n+1},\quad D_{\uparrow}v(t,x)(-1,-f(t,x,u))\leq\ell(t,x,u);$$

(v) *there exists a set* $C \subset]0, +\infty[$, *with* $\mu_{\mathcal{L}}(]0, +\infty[\backslash C)) = 0$, *such that for any* $(t, x) \in \operatorname{dom} v \cap (C \times \operatorname{int} \Omega)$

$$\forall u \in \mathbb{B} \subset \mathbb{R}^{n+1}, \quad -\ell(t, x, u) \leq D_{\downarrow} v(t, x)(1, f(t, x, u));$$

(vi) *if* $\operatorname{dom} v \neq \emptyset$ *then* $t \rightsquigarrow \operatorname{epi} v(t, .)$ *is of LBV. Moreover, if*

$$\exists \varphi \in L^1(\mathbb{R}_0^+; \mathbb{R}_0^+) : H^*(t, x, q) \geq -\varphi(t) \quad \forall t > 0, \forall x \in \mathbb{R}^n, \forall q \in \mathbb{R}^n, \tag{4.19}$$

then $t \rightsquigarrow \operatorname{epi} v(t, .)$ *is l.a.c.*

We need the following result.

Lemma 4.2.5 *Let* Ξ *be the set-valued map defined by*

$$\Xi(s, (x, \beta)) := \left\{ (f(t, x, u), -\ell(t, x, u) - r) \;\middle|\; \begin{array}{l} u \in \mathbb{B} \subset \mathbb{R}^{n+1}, \\ r \in [0, c(t)(1 + |x|) - \ell(t, x, u)] \end{array} \right\}.$$

for all $\beta \in \mathbb{R}$. *Then, for any* $R > 0$ *and* $T > 0$, *there exist* $\mu \in L^1([0, T]; \mathbb{R}_0^+)$ *and a set-valued map* $\Phi : [0, T] \times \mathbb{R}^{n+1} \rightsquigarrow \mathbb{R}^{n+1}$ *satisfying*

$$\operatorname{Ker}(\bar{t}, (\bar{x}, \bar{\beta}), \Phi) = \operatorname{Ker}(\bar{t}, (\bar{x}, \bar{\beta}), \Xi) \quad \forall \bar{t} \in [0, T], \forall (\bar{x}, \beta) \in B(0, R), \tag{4.20}$$

and:

(i) Φ *has nonempty convex closed values;*
(ii) $\Phi(., x)$ *is measurable for any* $x \in \mathbb{R}^n$;
(iii) $\Phi(t, .)$ *is continuous;*
(iv) $\sup_{q \in \Phi(t, (x, \beta))} |q| \leq \mu(t)$ *for a.e.* $t \in [0, T]$ *and all* $x \in \mathbb{R}^n$, $\beta \in \mathbb{R}$.

Proof We first show that Ξ enjoy the properties (i)–(iii). Notice that, in virtue of Lemma 4.2.3:(iii), (iv), the set-valued map

$$x \rightsquigarrow \{(f(t, x, u), \ell(t, x, u)) \,|\, u \in \mathbb{B} \subset \mathbb{R}^{n+1}\}$$

is continuous for a.e. $t \geq 0$ and closed valued for all $t \geq 0$.

Now, consider $t \geq 0$ and $x \in \mathbb{R}^n$ for which it is closed valued. Let

$$(f(t, x, u_k), -\ell(t, x, u_k) - r_k) \to (a, b) \in \mathbb{R}^n \times \mathbb{R}$$

with $u_k \in \mathbb{B}$ and $r_k \in [0, c(t)(1 + |x|) - \ell(t, x, u_k)]$ for all $k \in \mathbb{N}$. Since $(\ell(t, x, u_k))_k$ is bounded we deduce that $(r_k)_k$ is bounded. So, we may assume that $r_k \to r \geq 0$. Then $(f(t, x, u_k), \ell(t, x, u_k)) \to (a, -b - r)$, and, by closedness, there exists $u \in \mathbb{B}$ such that $a = f(t, x, u)$ and $-b - r = \ell(t, x, u)$. This proves that $\Xi(t, (x, \beta))$ is closed for any $\beta \in \mathbb{R}$.

Now, let $t \in \mathbb{R}_0^+$ be such that $x \rightsquigarrow \{(f(t, x, u), \ell(t, x, u)) \,|\, u \in \mathbb{B}\}$ is continuous. Then $(x, \beta) \rightsquigarrow \Xi_1(t, (x, \beta)) := \{(f(t, x, u), -\ell(t, x, u)) \,|\, u \in \mathbb{B}\}$ and

$(x, \beta) \rightsquigarrow \Xi_2(t, (x, \beta)) := \{(f(t, x, u), -c(t)(1 + |x|)) \mid u \in \mathbb{B}\}$ are continuous. Thus $(x, \beta) \rightsquigarrow \Xi_1(t, (x, \beta)) \cup \Xi_2(t, (x, \beta))$ is continuous, and it follows that $\Gamma : (x, \beta) \rightsquigarrow \text{cl co}(\Xi_1(t, (x, \beta)) \cup \Xi_2(t, (x, \beta)))$ is continuous too. Since $\Xi(t, (x, \beta)) = \Gamma(x, \beta)$, we deduce that $\Xi(t, (x, \beta))$ is convex and $\Xi(t, .)$ is continuous.

Let us define for any $R > 0$, $T > 0$, and $M > 0$, the set-valued map $\Phi : [0, T] \times \mathbb{R}^{n+1} \rightsquigarrow \mathbb{R}^{n+1}$ by

$$\Phi(t, X) := \begin{cases} \Xi(t, X) & (t, X) \in [0, T] \times B(0, M), \\ \Xi(t, \Pi_{B(0,M)}(X)) & (t, X) \in [0, T] \times (\mathbb{R}^{n+1} \backslash B(0, M)), \end{cases} \tag{4.21}$$

where $\Pi_{B(0,M)}(.)$ is the projection operator defined in (1.4).
(i)–(iii) It follows from the definition of Ξ and since Φ also enjoys the desiderate properties.
(iv): Let $(\bar{t}, \bar{x}) \in [0, T] \times \mathbb{R}^n$. From Gronwall's Lemma and our assumptions, any locally absolutely continuous trajectory $x : [0, T] \to \mathbb{R}^n$ solving the differential equation in (4.16) and starting from $\bar{x}$ at time $\bar{t}$ satisfies $1 + |x(t)| \le (1 + |\bar{x}|)\, e^{\int_{\bar{t}}^{t} c(s)\, ds}$ for all $t \in [\bar{t}, T]$. In particular, feasible trajectories starting at the same initial condition are uniformly bounded on every finite time interval. Now, fixing $R > 0$ and defining

$$\tilde{\mu}(t) := (1 + R)\, c(t) e^{\int_0^t c(s)\, ds}$$

it follows that $\tilde{\mu} \in L^1([0, T]; \mathbb{R}_0^+)$. Furthermore, for any $T > 0$, any $(\bar{t}, \bar{x}) \in [0, T] \times (\Omega \cap B(0, R))$, and any feasible trajectory-control pair $(x(.), u(.))$ on $I = [\bar{t}, T]$, with $x(\bar{t}) = \bar{x}$, we have for a.e. $t \in [\bar{t}, T]$

$$|f(t, x(t), u(t))|\, , |\ell(t, x(t), u(t))| \le \tilde{\mu}(t).$$

Now, fix $R > 0$, $T > 0$, and put $M := R + 2\int_0^T \tilde{\mu}(s)\, ds$. We have from the definition (4.21) that

$$\sup_{\substack{v \in \Phi(t,X) \\ X \in \mathbb{R}^{n+1}}} |v| \le 2\tilde{\mu}(t) =: \mu(t) \quad \text{a.e. } t \in [0, T].$$

Finally, from the construction of Ξ given in (4.21), we get immediately that $X : [\bar{t}, T] \to \mathbb{R}^{n+1}$, with $X(\bar{t}) \in B(0, R)$, is a Φ-trajectory if and only if it is a Ξ-trajectory on $[\bar{t}, T]$. So, it follows (4.20). □

Proof of Proposition 4.2.4 Let $(t_0, x_0) \in \text{dom}\, v$. Consider a minimizing sequence of feasible trajectory-control pairs (x_i, u_i) satisfying $x_i(t_0) = x_0$ for all $i \in \mathbb{N}$. Notice that

$$\lim_{T \to +\infty} \int_{t_0}^{T} \ell(t, x_i(t), u_i(t))dt \text{ exists finite for any } i \in \mathbb{N} \text{ and } \to_{i \to +\infty} v(t_0, x_0).$$

By Gronwall Lemma, for every $T > t_0$, the restrictions of x_i to $[t_0, T]$ are equibounded. Using Ascoli-Arzela Theorem, there exists a subsequence $(x_{i_k})_k$ and a continuous function $\bar{x} : \mathbb{R}^+_{t_0} \to \mathbb{R}^n$ such that for every $T > t_0$

$$x_{i_k} \to \bar{x} \quad \text{in } C([t_0, T]; \mathbb{R}^n).$$

In particular, $\bar{x}(.) \subset \Omega$. By the Dunford-Pettis Theorem, taking a subsequence and keeping the same notation, we may assume that there exist two locally integrable functions $y : \mathbb{R}^+_{t_0} \to \mathbb{R}^n$, $z : \mathbb{R}^+_{t_0} \to \mathbb{R}$ such that for every $T > t_0$

$$(x'_{i_k}(.), \ell(., x_{i_k}(.), u_{i_k}(.)) \rightharpoonup (y, z) \quad \text{in } L^1([t_0, T]; \mathbb{R}^n \times \mathbb{R}).$$

Passing to the limit, we get

$$\bar{x}(t) = x_0 + \int_{t_0}^{t} y(s)ds \quad \forall t \in \mathbb{R}^+_{t_0}.$$

Hence $\bar{x}$ is locally absolutely continuous and, by the Lebesgue Theorem, $\bar{x}'(t) = y(t)$ a.e. $t \in \mathbb{R}^+_{t_0}$. Notice that for any $S \geq T > t_0$ and any trajectory-control pair (x, u)

$$\begin{aligned} &\int_{t_0}^{S} \ell\,(t, x(t), u(t)) + \varphi(t)(1 + |x(t)|)dt \\ &\geq \int_{t_0}^{T} \ell\,(t, x(t), u(t)) + \varphi(t)(1 + |x(t)|)dt. \end{aligned} \tag{4.22}$$

Notice that, in virtue of Lemma 4.2.3:(ii), the function $t \mapsto \ell\,(t, x(t), u(t)) + \varphi(t)(1 + |x(t)|)$ is nonnegative real-valued. Furthermore, we have also that for any $S \geq T > t_0$

$$\begin{aligned} \int_{T}^{S} |\varphi(t)(1 + |x(t)|)|dt &\leq \int_{T}^{S} \varphi(t)(1 + (1 + |x_0|)e^{\int_0^t c(s)ds})dt \\ &\leq \int_{0}^{+\infty} \varphi(t)(1 + (1 + |x_0|)e^{\int_0^t c(s)ds})dt < +\infty. \end{aligned} \tag{4.23}$$

Now, fix $\varepsilon > 0$. Then, from (4.22) and (4.23), there exists $\bar{T}_\varepsilon > t_0$ such that for all $S \geq T \geq \bar{T}_\varepsilon$ and all $k \in \mathbb{N}$

$$\int_{t_0}^{S} \ell\left(t, x_{i_k}(t), u_{i_k}(t)\right) dt \geq \int_{t_0}^{T} \ell\left(t, x_{i_k}(t), u_{i_k}(t)\right) dt - \varepsilon.$$

Taking the limit as $S \to +\infty$ we get

$$\lim_{S\to+\infty}\int_{t_0}^{S}\ell\left(t,x_{i_k}(t),u_{i_k}(t)\right)dt \geq \int_{t_0}^{T}\ell\left(t,x_{i_k}(t),u_{i_k}(t)\right)dt-\varepsilon \tag{4.24}$$

and keeping the limit as $k\to+\infty$ in both side of the previous inequality

$$v\left(t_0,x_0\right)\geq\int_{t_0}^{T}z(t)dt-\varepsilon \quad \forall T\geq\bar{T}_{\varepsilon}. \tag{4.25}$$

Passing to the upper limit as $T\to+\infty$ and since $\varepsilon>0$ is arbitrary, we conclude that

$$v\left(t_0,x_0\right)\geq\limsup_{T\to+\infty}\int_{t_0}^{T}z(t)dt. \tag{4.26}$$

Now, fix $T>t_0$, $\varepsilon>0$, and observe that, from Lemma 4.2.3:(iv), for a.e. $t\in[t_0,T]$ and for all large $k\in\mathbb{N}$

$$\begin{aligned}(x'_{i_k}(t),\ell(t,x_{i_k}(t),u_{i_k}(t)))&\in\{(f(t,x_{i_k}(t),u),\ell(t,x_{i_k}(t),u)+r)|u\in\mathbb{B},r\geq0\}\\&\subset\{(f(t,\bar{x}(t),u),\ell(t,\bar{x}(t),u)+r)|u\in\mathbb{B},r\geq0\}+\varepsilon\mathbb{B}.\end{aligned}$$

The set in the above inclusion is closed and convex by Lemma 4.2.3:(v). Since the restrictions

$$(x'_{i_k}(.),\ell(.,x_{i_k}(.),u_{i_k}(.)))|_{[t_0,T]}\rightharpoonup(\bar{x}',z)|_{[t_0,T]} \quad \text{in } L^1\left([t_0,T];\mathbb{R}^n\times\mathbb{R}\right)$$

from the Mazur Theorem we deduce that

$$\left(\bar{x}'(t),z(t)\right)\in\{(f(t,\bar{x}(t),u),\ell(t,\bar{x}(t),u)+r)|u\in\mathbb{B},r\geq0\}+\varepsilon\mathbb{B} \quad \text{a.e. } t\in[t_0,T].$$

By the arbitrariness of $\varepsilon>0$ and $T>0$,

$$\left(\bar{x}'(t),z(t)\right)\in\{(f(t,\bar{x}(t),u),\ell(t,\bar{x}(t),u)+r)|u\in\mathbb{B},r\geq0\}$$

for a.e. $t\in[t_0,+\infty[$. By the Measurable Selection Theorem there exist a control $\bar{u}(.)$ and a measurable function $r:\mathbb{R}^+_{t_0}\to\mathbb{R}^+_0$ such that

$$\bar{x}'(t)=f(t,\bar{x}(t),\bar{u}(t)), \quad z(t)=\ell(t,\bar{x}(t),\bar{u}(t))+r(t). \tag{4.27}$$

Hence, from (4.22) and (4.23), we get

$$\lim_{S\to+\infty}\int_{t_0}^{S}\ell(t,\bar{x}(t),\bar{u}(t))dt>-\infty. \tag{4.28}$$

From (4.26), (4.27), and (4.28), we finally conclude that

$$\lim_{T\to+\infty}\int_{t_0}^{T} \ell(t,\bar{x}(t),\bar{u}(t))dt$$
$$\geq v\,(t_0,x_0)$$
$$\geq \limsup_{T\to+\infty}\int_{t_0}^{T} z(t)dt \geq \limsup_{T\to+\infty}\int_{t_0}^{T} \ell(t,\bar{x}(t),\bar{u}(t))dt.$$

So, (i) holds true.

Next, we show (ii). Consider a sequence

$$(t_0^i, x_0^i) \to_{i\to+\infty} (t_0, x_0).$$

We want to show $\liminf_{i\to+\infty} v(t_0^i, x_0^i) \geq v\,(t_0, x_0)$. It is enough to consider the case $(v(t_0^i, x_0^i))_{i\in\mathbb{N}}$ is bounded. Let a subsequence

$$\lim_{k\to+\infty} v(t_0^{i_k}, x_0^{i_k}) = \liminf_{i\to+\infty} v(t_0^i, x_0^i)$$

and (x_{i_k}, u_{i_k}) be the corresponding optimal trajectory-control pairs wrt $(t_0^{i_k}, x_0^{i_k})$. We extend $(x_{i_k}(.), \ell(., x_{i_k}(.), u_{i_k}(.)))$. on $[0, t_0^{i_k}[$ by setting

$$x_{i_k}(s) = x_{i_k}(t_0^{i_k}) \text{ and } \ell(s, x_{i_k}(s), u_{i_k}(s)) = 0 \quad \forall s \in [0, t_0^{i_k}[.$$

Using the same arguments as in the previous proof, without labeling and maintaining the same notation, we extract a subsequence x_{i_k} converging almost uniformly to a locally absolutely continuous function $\bar{x} : \mathbb{R}_0^+ \to \mathbb{R}^n$, satisfying $\bar{x}\,(t_0) = x_0$, $\bar{x}(.) \subset \Omega$, and such that for every large $T > t_0$ the restrictions of $(x'_{i_k}(.), \ell(., x_{i_k}(.), u_{i_k}(.)))$ to $[0, T]$ converge weakly in $L^1([0, T]; \mathbb{R}^n \times \mathbb{R})$ to $(\bar{x}', z)|_{[0,T]}$, where $z : \mathbb{R}_0^+ \to \mathbb{R}$ is a locally integrable function satisfying (4.27) for a suitable control $\bar{u}(.)$ and a measurable function $r : \mathbb{R}_0^+ \to \mathbb{R}_0^+$. Notice $\bar{x}(t) = x_0$, $z(t) = 0$, for all $t \in [0, t_0]$. From (4.24)–(4.27), we get for all large $T > 0$

$$\lim_{k\to+\infty} v(t_0^{i_k}, x_0^{i_k}) = \lim_{k\to+\infty}\liminf_{S\to+\infty}\int_{t_0^{i_k}}^{S} \ell(t, x_{i_k}(t), u_{i_k}(t))dt \geq \int_{t_0}^{T} z(t)dt$$

that, using the same arguments as in the proof of part (i), imply

$$\lim_{k\to+\infty} v(t_0^{i_k}, x_0^{i_k}) \geq v(t_0, x_0).$$

Now, let $x \in \Omega$. If $v(0, x) = +\infty$ then, since v is lower semicontinuous, (4.17) holds true. Suppose next that $(0, x) \in \text{dom}\, v$. Consider an optimal trajectory-control pair $(\bar{x}(.), \bar{u}(.))$ at $(0, x)$. Then, by the dynamic programming principle, for all $t \geq 0$

$$v(t, \bar{x}(t)) = v(0, x) - \int_0^t \ell(s, \bar{x}(s), \bar{u}(s))\, ds.$$

So, $\lim_{t \to 0+} v(t, \bar{x}(t)) = v(0, x)$. The lower semicontinuity of v ends the proof of (ii).

Consider the set-valued map Φ provided by Lemma 4.2.5.

We show (iii). Let $j \in \mathbb{N}^+$. From Lemma 2.2.5 applied to the set-valued map Φ, there exists a set $C_j \subset [0, j]$, with $\mu_{\mathscr{L}}(C_j) = 0$, such that for any $(t_0, x_0) \in (([0, j] \backslash C_j) \times \Omega) \cap \operatorname{dom} v$ and any optimal trajectory-control pair $(\bar{x}(.), \bar{u}(.))$ at (t_0, x_0),

$$\emptyset \neq \operatorname*{Lim\,sup}_{\xi \to t_0+} \left\{ \frac{1}{\xi - t_0} \left(\bar{x}(\xi) - x_0, - \int_{t_0}^{\xi} \ell(s, \bar{x}(s), \bar{u}(s))\, ds \right) \right\} \subset \Phi(t_0, x_0, \beta_0) \tag{4.29}$$

for all $\beta_0 \in \mathbb{R}$. Furthermore, by the dynamic programming principle, for all $t \geq t_0$

$$v(t, \bar{x}(t)) - v(t_0, x_0) = - \int_{t_0}^{t} \ell(s, \bar{x}(s), \bar{u}(s))\, ds.$$

So, dividing by $t - t_0$ this equality, passing to the lower limit as $t \to t_0+$, and using (4.29), we get (4.18). Then (iii) follows setting $C = \cup_{j \in \mathbb{N}^+} C_j$.

Next, we prove (iv). Let $j \in \mathbb{N}^+$. Recalling Lemmata 4.2.3, 4.2.5, and 2.2.5, to the set-valued map $[1/j, j] \times \mathbb{R}^n \times \mathbb{R} \ni (s, x, \beta) \rightsquigarrow -\Phi(j - s, x, \beta) \in \mathbb{R}^n \times \mathbb{R}$, and the Measurable Selection Theorem we have that there exists a subset $C_j \subset [1/j, j]$, with $\mu_{\mathscr{L}}(C_j) = 0$, such that for any $(t_0, x_0) \in (]1/j, j] \backslash C_j) \times \operatorname{int} \Omega$ and any $u_0 \in \mathbb{B}$ we can find $t_1 \in [1/j, t_0[$ and a trajectory-control pair $((\xi, \beta), (u, r))(.)$ satisfying

$$\begin{cases} (\xi, \beta)'(t) = (f(t, \xi(t), u(t)), -\ell(t, \xi(t), u(t)) - r(t)) & \text{a.e. } t \in [t_1, t_0], \\ (u, r)(t) \in \mathbb{B} \times [0, c(t)(1 + |\xi(t)|) - \ell(t, \xi(t), u(t))] & \text{a.e. } t \in [t_1, t_0], \\ (\xi, \beta)(t_0) = (x_0, 0), \\ (\xi, \beta)'(t_0) = (f(t_0, x_0, u_0), -\ell(t_0, x_0, u_0)), \\ \xi([t_1, t_0]) \subset \Omega. \end{cases}$$

Hence, if $(t_0, x_0) \in \operatorname{dom} v$, by the dynamic programming principle it follows that

$$\frac{v(s, \xi(s)) - v(t_0, x_0)}{t_0 - s} \leq \frac{1}{t_0 - s} (\beta(s) - \beta(t_0))$$

for all $s \in [t_1, t_0]$. Passing to the lower limit as $s \to t_0-$ and using the lower semicontinuity of v, we conclude $D_{\uparrow} v(t_0, x_0)(-1, -f(t_0, x_0, u_0)) \leq \ell(t_0, x_0, u_0)$. Since $u_0 \in \mathbb{B}$ is arbitrary, the statement (iv) follows with $C =]0, +\infty[\backslash \cup_{j \in \mathbb{N}} C_j$.

The statement (v) holds as well arguing in a similar way as in (iv).

Next, we show (vi). Assume that $\operatorname{dom} v \neq \emptyset$ and denote

$$\mathscr{F}(t, x) := f(t, x, \mathbb{B}).$$

Notice that the value function v is bounded from the below and since it is lower semicontinuous, $t \rightsquigarrow \operatorname{epi} v(t, .)$ takes closed images. Let $(\bar{t}, \bar{x}) \in \operatorname{dom} v$. Then, by the dynamic programming principle, it follows that the set-valued map $t \rightsquigarrow \operatorname{epi} v(t, .)$ takes nonempty values on $[\bar{t}, +\infty[$. If $\bar{t} > 0$, consider $\tau \in [0, \bar{t}[$. Set $E(t) = \Omega$ for all $t \geq 0$. Now, from Assumption 4.2.2 it follows that $(\{1\} \times (-\mathscr{F}(t, x)) \cap T_\Omega(t, x) \neq \emptyset$ for a.e. $t \in]\tau, \bar{t}]$ and all $x \in \Omega$. Hence, the viability Theorem 2.2.4 applied to the set-valued map $(s, y) \rightsquigarrow \Phi(s, y) = -\mathscr{F}(\bar{t} - s, y)$ and the constant tube $E(.) \equiv \Omega$, and the Measurable Selection Theorem, imply that there exists a feasible trajectory-control pair $(\tilde{x}(.), \tilde{u}(.))$ on $I = [\tau, \bar{t}]$ satisfying $\tilde{x}(\bar{t}) = \bar{x}$. So, applying again the dynamic programming principle and since $\tau \in [0, \bar{t}[$ is arbitrary, it follows that $t \rightsquigarrow \operatorname{epi} v(t, .)$ takes nonempty values on $[0, \bar{t}]$.

Now, fix a real interval $[a, b]$ and $\mathscr{K} \subset \mathbb{R}^{n+1}$ a nonempty compact subset. Let any $0 \leq t_1 \leq t_0$ and $(x_1, v_1) \in \operatorname{epi} v(t_1, \cdot) \cap \mathscr{K}$. Consider an optimal trajectory-control pair $(\bar{x}(.), \bar{u}(.))$ at (t_1, x_1). Then, from the dynamic programming principle,

$$
\begin{aligned}
& v(t_1, x_1) + \int_{t_1}^{t_0} \varphi(s)(1 + |\bar{x}(s)|)ds \\
& = \int_{t_1}^{+\infty} \ell(s, \bar{x}(s), \bar{u}(s))\, ds + \int_{t_1}^{t_0} \varphi(s)(1 + |\bar{x}(s)|)ds \\
& \geq \int_{t_0}^{+\infty} \ell(s, \bar{x}(s), \bar{u}(s))\, ds = v(t_0, \bar{x}(t_0)).
\end{aligned}
$$

Since $v_1 \geq v(t_1, x_1)$ we get $(\bar{x}(t_0), v_1 + \int_{t_1}^{t_0} \varphi(s)(1 + |\bar{x}(s)|)ds) \in \operatorname{epi} v(t_0, \cdot)$. Recalling (4.23), we deduce that

$$
\begin{aligned}
(x_1, v_1) \in \operatorname{epi} v(t_0, \cdot) & \\
& + (\int_{t_1}^{t_0} (1 + |x_1|)(1 + e^{\int_{t_1}^{s} c(t)dt})ds \\
& + \int_{t_1}^{t_0} \varphi(s)(1 + (1 + |x_1|)e^{\int_{t_1}^{s} c(t)dt} ds)\mathbb{B}.
\end{aligned}
$$

On the other hand, let $(x_0, v_0) \in \operatorname{epi} v(t_0, \cdot) \cap \mathscr{K}$. Applying again Lemma 4.2.5, Theorem 2.2.4, and the Measurable Selection Theorem, we deduce that there exists a feasible trajectory-control pair $(\tilde{x}(.), \tilde{u}(.))$ on $I = [t_1, t_0]$ satisfying $\tilde{x}(t_0) = x_0$. So, by the dynamic programming principle, we get

$$
\begin{aligned}
v(t_1, \tilde{x}(t_1)) & \leq v(t_0, x_0) + \int_{t_1}^{t_0} \ell(s, \tilde{x}(s), \tilde{u}(s))\, ds \\
& \leq v_0 + \int_{t_1}^{t_0} c(s)(1 + |x_0|)e^{\int_{t_1}^{s} c(t)dt} ds
\end{aligned}
$$

i.e.,

$$(\tilde{x}(t_1), v_0 + \int_{t_1}^{t_0} c(s)(1 + |x_0|)e^{\int_{t_1}^{s} c(t)dt} ds) \in \operatorname{epi} v(t_1, \cdot).$$

Taking into account

$$\begin{aligned}(x_0, v_0) = (\tilde{x}(t_1), v_0 + \int_{t_1}^{t_0} c(s)(1 + |x_0|)e^{\int_{t_1}^{s} c(t)dt} ds)\\ + (x_0 - \tilde{x}(t_1), - \int_{t_1}^{t_0} c(s)(1 + |x_0|)e^{\int_{t_1}^{s} c(t)dt} ds)\end{aligned}$$

we conclude

$$\begin{aligned}&(x_0, v_0) \in \operatorname{epi} v(t_1, \cdot)\\ &\quad + \left(\int_{t_1}^{t_0} 4(1 + |x_0|)(1 + e^{\int_{t_1}^{s} c(t)dt} \right) ds + \int_{t_1}^{t_0} c(s)(1 + |x_0|)e^{\int_{t_1}^{s} c(t)dt} ds)\mathbb{B}.\end{aligned}$$

Finally,

$$\begin{aligned}\sup_{\substack{x_1 \in \mathscr{K} \\ a=t_1<t_i<\ldots<t_{i+1}<t_m=b \\ \text{finite partition}}} &\sum_i \left(\int_{t_i}^{t_{i+1}} (1 + |x_1|)(1 + e^{\int_{t_i}^{s} c(t)dt} \right) ds\\ &+ \int_{t_i}^{t_{i+1}} \varphi(s)(1 + (1 + |x_1|)e^{\int_{t_i}^{s} c(t)dt}) ds\\ \leq \int_a^b (1 + R)(1 + e^{\int_a^s c(t)dt}) &+ \varphi(s)(1 + (1 + R)e^{\int_a^s c(t)dt}) ds \; < +\infty,\end{aligned}$$

$$\begin{aligned}\sup_{\substack{x_0 \in \mathscr{K} \\ a=t_1<t_i<\ldots<t_{i+1}<t_m=b \\ \text{finite partition}}} &\sum_i \left(\int_{t_i}^{t_{i+1}} 4(1 + |x_0|)(1 + e^{\int_{t_i}^{s} c(t)dt} \right) ds\\ &+ \int_{t_i}^{t_{i+1}} c(s)(1 + |x_0|)e^{\int_{t_i}^{s} c(t)dt} ds)\\ \leq \int_a^b 4(1 + R)(1 + e^{\int_a^s c(t)dt}) &+ c(s)(1 + R)e^{\int_a^s c(t)dt} ds < +\infty,\end{aligned}$$

where $R := \sup_{k \in \mathscr{K}} |k|$. Therefore, $t \rightsquigarrow \operatorname{epi} v(t, .)$ is of LBV. Using the same arguments, the last statement in (vi) easily follows. □

4.3 Weak Solutions

When dealing with HJB equations in infinite horizon problems with state constraints, classical smooth solutions often fail to exist due to the inherent nonsmooth nature of the value function and the presence of constraints. This necessitates the development of a weak solution theory that can handle such irregularities while preserving the essential properties needed for optimal control applications. We introduce and analyze the concept of weak (epigraphical) solutions for a class of HJB equations, providing uniqueness results.

In this section, for any Hamiltonian $H : \mathbb{R}_0^+ \times \mathbb{R}^n \times \mathbb{R}^n \to \mathbb{R}$ satisfying Assumption 4.1.7 we denote by $H_{rep} : \mathbb{R}_0^+ \times \mathbb{R}^n \times \mathbb{R}^n \times \mathbb{R} \to \mathbb{R}$ the following function

$$H_{rep}(t, x, p, q) := \sup_{u \in \mathbb{B} \subset \mathbb{R}^{n+1}} \{\langle p, f(t, x, u)\rangle + q\ell(t, x, u)\}$$

where (f, ℓ) is the representation given by Theorem 4.1.8 associated to H.

Definition 4.3.1 We say that a lower semicontinuous function $V : \mathbb{R}_0^+ \times \Omega \to [-\infty, +\infty]$ is an *weak (epigraphical) solution* of the HJB equation

$$-\partial V_t(t, x) + H(t, x, -\nabla_x V(t, x)) = 0 \quad (t, x) \in]0, +\infty[\times \Omega \tag{4.30}$$

if there exists a set $C \subset]0, +\infty[$, with $\mu_{\mathscr{L}}(C) = 0$, such that for all $(t, x) \in \operatorname{dom} V \cap ((]0, +\infty[\backslash C) \times \operatorname{bdr} \Omega)$

$$-p_t + H_{rep}(t, x, -p_x, -q) \geq 0 \quad \forall (p_t, p_x, q) \in T_{\operatorname{epi} V}(t, x, V(t, x))^-,$$

and for all $(t, x) \in \operatorname{dom} V \cap ((]0, +\infty[\backslash C) \times \operatorname{int} \Omega)$

$$-p_t + H_{rep}(t, x, -p_x, -q) = 0 \quad \forall (p_t, p_x, q) \in T_{\operatorname{epi} V}(t, x, V(t, x))^-.$$

Remark 4.3.2 (i) If V is locally Lipschitz continuous, then Definition 4.3.1 is equivalent to the notion of *viscosity* solution in terms of subdifferentials for Hamiltonian arising from control problems:

$$-p_t + H(t, x, -p_x) \geq 0, \quad \forall (p_t, p_x) \in \partial_- V(t, x), \text{ for a.e. } t > 0, \forall x \in \Omega,$$
$$-p_t + H(t, x, -p_x) \leq 0, \quad \forall (p_t, p_x) \in \partial_+ V(t, x), \text{ for a.e. } t > 0, \forall x \in \operatorname{int} \Omega.$$

(ii) From the properties of normal and tangent cones, it follows that a function V is a weak epigraphical solution if and only if there exists a set $C \subset]0, +\infty[$, with $\mu_{\mathscr{L}}(C) = 0$, such that for all $(t, x) \in \operatorname{dom} V \cap ((]0, +\infty[\backslash C) \times \operatorname{bdr} \Omega)$

$$-p_t + H(t, x, -p_x) \geq 0 \quad \forall (p_t, p_x, -1) \in T_{\operatorname{epi} V}(t, x, V(t, x))^-,$$
$$-p_t + \sup_{q \in D(t,x)} \langle q, -p_x\rangle \geq 0, \quad \forall (p_t, p_x, 0) \in T_{\operatorname{epi} V}(t, x, V(t, x))^-,$$

and for all $(t, x) \in \operatorname{dom} V \cap ((]0, +\infty[\backslash C) \times \operatorname{int} \Omega)$

$$-p_t + H(t, x, -p_x) = 0 \quad \forall (p_t, p_x, -1) \in T_{\operatorname{epi} V}(t, x, V(t, x))^-,$$
$$-p_t + \sup_{q \in D(t,x)} \langle q, -p_x \rangle = 0, \quad \forall (p_t, p_x, 0) \in T_{\operatorname{epi} V}(t, x, V(t, x))^-.$$

Definition 4.3.3 By $\mathcal{E}pi_{\mathrm{loc}}(\mathbb{R}_0^+ \times \Omega)$ we denote the family of all lower semicontinuous functions $V : \mathbb{R}_0^+ \times \Omega \to [-\infty, +\infty]$ such that

$$(t \rightsquigarrow \operatorname{epi} V(t, .)) \text{ is continuous and of LBV on } \mathbb{R}_0^+.$$

4.3.1 *Uniqueness of Solutions*

The uniqueness of weak solutions plays a crucial role in characterizing the value function via the HJB equation, yet it presents significant technical challenges, especially in infinite horizon settings. The presence of state constraints introduce additional complexity in establishing results compared to the classical smooth case. This section presents comprehensive uniqueness theorems for weak solutions under appropriate assumptions on the Hamiltonian structure and problem data.

Theorem 4.3.4 (Uniqueness) *Consider Assumptions 4.2.1 and 2.3.5, the latter satisfied with $F(t, x) = D(t, x)$. Suppose that for every $(t, x) \in \mathbb{R}_0^+ \times \Omega$:*

(i) $\lim_{T \to +\infty} \int_t^T H^*(s, \xi(s), \xi'(s))\, ds$ *exists for any* $\xi \in W_{\mathrm{loc}}^{1,1}(\mathbb{R}_t^+; \Omega)$ *such that* $\xi(t) = x$*;*
(ii) the infimum $\alpha(t, x)$ defined by

$$\inf \left\{ \lim_{T \to +\infty} \int_t^T H^*(s, \xi(s), \xi'(s))\, ds \mid \xi \in W_{\mathrm{loc}}^{1,1}(\mathbb{R}_t^+; \Omega), \xi(t) = x \right\}$$

is finite.

Then, if there exist

$$u, w \in \mathcal{E}pi_{\mathrm{loc}}(\mathbb{R}_0^+; \mathbb{R}^{n+1})$$

weak epigraphical solutions of the HJB equation with vanishing condition

$$\begin{cases} -\partial V_t(t, x) + H(t, x, -\nabla_x V(t, x)) = 0 & (t, x) \in]0, +\infty[\times \Omega, \\ \lim_{t \to +\infty} \sup_{x \in \Omega} |V(t, x)| = 0, \end{cases} \tag{4.31}$$

then $v = w$.

Corollary 4.3.5 *Consider the assumptions of Theorem 4.3.4 with the further condition* (4.19)*. If u, w are weak solutions of the HJB equation with vanishing condition (4.31) and $t \rightsquigarrow \operatorname{epi} u(t,.)$, $t \rightsquigarrow \operatorname{epi} w(t,.)$ are l.a.c., then*

$$u = v = w \quad (\text{on dom } v).$$

In what follows, we consider the representation associated with the Hamiltonian H

$$(f, \ell) : \mathbb{R}_0^+ \times \mathbb{R}^n \times \mathbb{B} \to \mathbb{R}^n \times \mathbb{R}$$

provided by Theorem 4.1.8.

Proof of Theorem 4.3.4 Consider two epigraphical weak solutions of the HJB equation, namely v and w, satisfying the vanishing condition in (4.31). It is sufficient to show that $u \le v \le w$. Fix $(t_0, x_0) \in]0, +\infty[\times\Omega$. By our assumptions, there exists $T > t_0$ such that

$$|w(t, y)| \le \varepsilon, \; |u(t, y)| \le \varepsilon \quad \forall t \ge T, \; \forall y \in \Omega. \tag{4.32}$$

We divide the proof into parts (A), (B), and (C).

(A): Wc show that

$$w(t_0, x_0) \ge v(t_0, x_0). \tag{4.33}$$

If $w(t_0, x_0) = +\infty$, then $w(t_0, x_0) \ge v(t_0, x_0)$. So, assume that $(t_0, x_0) \in \operatorname{dom} w$. In order to prove (4.33), we show the following

$$\begin{aligned} &\exists (\xi, \theta) : \mathbb{R}_{t_0}^+ \to \mathbb{R}^n \times \mathbb{R}^{n+1} \text{ solving (4.16):} \\ &w(t_0, x_0) \ge w(t, \xi(t)) + \int_{t_0}^t L(s, \xi(s), \theta(s))\, ds \quad \forall t \ge t_0. \end{aligned} \tag{4.34}$$

We postpone the proof of (4.34) and we assume temporarily it is valid. Using the vanishing condition and passing to the upper limit in (4.34) as $t \to +\infty$ yields, for every $(\xi(.), \theta(.)) \in \mathscr{U}_\Omega(t_0, x_0)$,

$$w(t_0, x_0) \ge \limsup_{t\to+\infty} \int_{t_0}^t L(s, \xi(s), \theta(s))\, ds.$$

In particular, it follows that

$$\begin{aligned} w(t_0, x_0) &\ge \inf \left\{ \limsup_{t\to+\infty} \int_{t_0}^t H^*(s, \xi(s), \xi'(s))\, ds \;\middle|\; \begin{array}{l} \xi \in W^{1,1}_{\text{loc}}(\mathbb{R}_{t_0}^+; \Omega), \\ \xi(t_0) = x_0 \end{array} \right\} \\ &= \alpha(t_0, x_0). \end{aligned} \tag{4.35}$$

By our assumptions, $\alpha(t_0, x_0) > -\infty$. Fix $\varepsilon > 0$ and consider a trajectory $\xi \in W^{1,1}_{\text{loc}}(\mathbb{R}_{t_0}^+; \mathbb{R}^n)$ with $\xi(t_0) = x_0$ and $\xi(.) \subset \Omega$ satisfying

$$\int_{t_0}^{+\infty} H^*(s, \xi(s), \xi'(s))ds < \alpha(t_0, x_0) + \varepsilon.$$

We have that $(\xi'(s), z'(s)) \in \operatorname{graph} H^*(s, \xi(s), .)$ for a.e. $s \geq t_0$, where we put $z(s) := \int_t^s H^*(\tau, \xi(\tau), \xi'(\tau))\, d\tau$ for all $s \geq t_0$. Applying now Theorem 4.1.8:(v)' and the Measurable Selection Theorem, we have that there exists a measurable function $\theta : \mathbb{R}^+_{t_0} \to \mathbb{B} \subset \mathbb{R}^{n+1}$ such that

$$(\xi'(s), z'(s)) = (f(s, \xi(s), \theta(s)), \ell(s, \xi(s), \theta(s)))$$

for a.e. $s \geq t_0$. So, for all $t \geq t_0$

$$\begin{aligned} &\int_{t_0}^{t} H^*(s, \xi(s), \xi'(s))\, ds \\ &= \int_{t_0}^{t} z'(s)\, ds = \int_{t_0}^{t} \ell(s, \xi(s), \theta(s))\, ds, \end{aligned}$$

that imply

$$\int_{t_0}^{+\infty} H^*(s, \xi(s), \xi'(s))\, ds \geq v(t_0, x_0).$$

We get $\alpha(t_0, x_0) + \varepsilon > v(t_0, x_0)$. Since ε is arbitrary, we have $\alpha(t_0, x_0) \geq v(t_0, x_0)$. Recalling (4.35), it follows the inequality in (4.33).

To conclude the proof of (A), we have only to show (4.34). Since w is a weak epigraphical solution of the HJB, applying the representation result Theorem 4.1.8 there exists a set $C \subset \mathbb{R}^+_0$ with $\mu_{\mathcal{L}}(C) = 0$ such that for all $(t, x) \in \operatorname{dom} w \cap ((\mathbb{R}^+_0 \backslash C) \times \Omega)$

$$\begin{aligned} &- r + \sup\left\{\langle f(t, x, \theta), -p\rangle + q\ell(t, x, \theta) \,|\, \theta \in \mathbb{B} \subset \mathbb{R}^{n+1}\right\} \geq 0 \\ &\forall\, (r, p, q) \in T_{\text{epi }w}(t, x, w(t, x))^-. \end{aligned} \tag{4.36}$$

Hence, from (4.36) and the Separation Theorem, we deduce that

$$(\{1\} \times \Phi(t, x, z)) \cap \operatorname{cl}\operatorname{co} T_{\text{epi}\, w}(t, x, w(t, x)) \neq \emptyset \tag{4.37}$$

for all $(t, x) \in \operatorname{dom} w \cap ((\mathbb{R}^+_0 \backslash C) \times \Omega)$ and $z \in \mathbb{R}$, where Φ is the set-valued map provided by Lemma 4.2.5. Applying Theorem 2.2.4 with $E(t) = \operatorname{epi} w(t, .)$, we deduce that there exists an absolutely continuous trajectory $X_0(.) = (\xi_0(.), z_0(.))$ solving

$$\begin{cases} X'(t) \in \Phi(t, X(t)) & \text{a.e. } t \in [t_0, t_0 + 1], \\ \xi(t_0) = x_0, \\ z(t_0) = w(t_0, x_0), \\ \xi(t) \in \Omega,\ z(t) \geq w(t, \xi(t)) & \forall t \in [t_0, t_0 + 1]. \end{cases} \tag{4.38}$$

We claim that for any $j \in \mathbb{N}^+$ the trajectory X_0 admits an extension on the interval $[t_0, t_0 + j]$ to a trajectory X_j satisfying (4.38) on $[t_0, t_0 + j]$. We proceed by the induction argument on $j \in \mathbb{N}$. Let $j \in \mathbb{N}$ and suppose that $X_j(.) = (\xi_j(.), z_j(.))$ satisfies the claim. Then, using (4.37) and applying again Theorem 2.2.4 on the time interval $[t_0 + j, t_0 + j + 1]$, we can find a trajectory $X(.) = (\xi(.), z(.))$ satisfying

$$\begin{cases} X'(t) \in \Phi(t, X(t)) & \text{a.e. } t \in [t_0 + j, t_0 + j + 1], \\ \xi(t_0 + j) = \xi_j(t_0 + j), & \\ z(t_0 + j) = z_j(t_0 + j), & \\ \xi(t) \in \Omega,\ z(t) \geq w(t, \xi(t)) & \forall t \in [t_0 + j, t_0 + j + 1]. \end{cases}$$

Putting $X_{j+1}(t) = (\xi_j(t), z_j(t))$ if $t \in [t_0, t_0 + j]$ and $X_{j+1}(t) = (\xi(t), z(t))$ if $t \in]t_0 + j, t_0 + j + 1]$, we deduce that $X_{j+1}(.)$ satisfies our claim. Now, consider the trajectory $X(t) = (\xi(t), z(t))$ given by $X(t) = X_j(t)$ if $t \in [t_0 + j, t_0 + j + 1]$. By the Measurable Selection Theorem, there exist two measurable functions $\theta(.)$ and $r(.)$, with $\theta(t) \in \mathbb{B} \subset \mathbb{R}^{n+1}$ and $r(t) \in [0, c(t)(1 + |\xi(t)|) - \ell(t, \xi(t), \theta(t))]$ for a.e. $t \geq t_0$, such that

$$z(t) = w(t_0, x_0) - \int_{t_0}^{t} \ell(t, \xi(s), \theta(s))\, ds - \int_{t_0}^{t} r(s)\, ds \geq w(t, \xi(t))$$

for all $t \geq t_0$. Hence, inequality in (4.34) immediately follows.

(B): We show

$$u(t_0, x_0) \leq v(t_0, x_0).$$

If $v(t_0, x_0) = +\infty$, then $v(t_0, x_0) \geq u(t_0, x_0)$. So, let us assume that $(t_0, x_0) \in \operatorname{dom} v$. Fix $\varepsilon > 0$. Let $(\bar{\xi}(.), \bar{\theta}(.))$ be an optimal trajectory-control pair at (t_0, x_0) for v and consider $s_i \to +\infty$ with $(s_i)_{i \in \mathbb{N}} \subset]T, +\infty[$. Put $\bar{X}(.) = (\bar{\xi}(.), \bar{z}(.))$ where $\bar{z}(t) = -\int_{t_0}^{t} \ell(t, \bar{\xi}(s), \bar{\theta}(s))\, ds$. Consider now for any $i \in \mathbb{N}$ the construction $\widehat{F}_i(.,.) : [t_0, s_i] \times \mathbb{R}^n \to \mathbb{R}^n$ given in (4.21) associated to

$$F(s, (x, \beta)) := \Big\{ (f(t, x, \theta), \ell(t, x, \theta)) \mid \theta \in \mathbb{B} \subset \mathbb{R}^{n+1} \Big\}.$$

From Remark 4.1.13:*(iii)*, we have that for any $i \in \mathbb{N}$ there exists $R_i > 0$ such that for every $(t, x) \in [t_0, s_i] \times \mathbb{R}^n$

$$\forall \phi \in \widehat{F}_i(t, x),\ |\phi| \leq 5(1 + R_i)(\varphi(t) + \tilde{\varphi}(t) + \tilde{c}(t)),$$

$$\widehat{F}_i(t, x) \subset F(t, x).$$

Thus, the set-valued maps

$$i \in \mathbb{N}, \quad F_i(t,x) := \begin{cases} \widehat{F}_i(t,x) & t \in [t_0, s_i],\, x \in \mathbb{R}^n, \\ \widehat{F}_i(s_i,x) & t > s_i,\, x \in \mathbb{R}^n, \end{cases}$$

together with Assumption 2.3.5, satisfy the hypothesis of Theorem 2.3.6. We can deduce that for any $i \in \mathbb{N}$ there exists a trajectory $X_i(.) = (\xi_i(.), z_i(.))$ solving

$$\begin{cases} X_i'(t) \in F_i(t, X_i(t)) & \text{a.e. } t \in [t_0, s_i], \\ X_i(s_i) = (\bar{\xi}(s_i), \bar{z}(s_i)), \\ \xi_i(t) \in \operatorname{int}\Omega & \forall t \in [t_0, s_i[, \end{cases}$$

and

$$\lim_{i\to+\infty} \sup\{|X_i(s) - \bar{X}(s)| \mid s \in [t_0, s_i]\} = 0,$$

where, with a slight abuse of notation,

$$F_i(t, X = (x,z)) := F_i(t,x).$$

Hence, by the Measurable Selection Theorem, for any $i \in \mathbb{N}$ there exists a measurable selection $\theta_i(t) \in \mathbb{B} \subset \mathbb{R}^{n+1}$ such that $(\xi_i(.), \theta_i(.))$ satisfies

$$\begin{cases} \xi_i'(t) = f(t, \xi_i(t), \theta_i(t)) & \text{a.e. } t \in [t_0, s_i], \\ \xi_i(s_i) = \bar{\xi}(s_i), \\ \xi_i(t) \in \operatorname{int}\Omega & \forall t \in [t_0, s_i[, \end{cases}$$

and

$$\lim_{i\to+\infty} \xi_i(t_0) = \bar{\xi}(t_0), \tag{4.39}$$

$$\lim_{i\to+\infty} \int_{t_0}^{s_i} \ell(t, \bar{\xi}_i(s), \bar{\theta}_i(s))\, ds = \int_{t_0}^{+\infty} \ell(t, \bar{\xi}(s), \bar{\theta}(s))\, ds. \tag{4.40}$$

Now, fix $i \in \mathbb{N}$ and consider $(\tau_j)_j \subset]T, s_i[$ with $\tau_j \to s_i$. Note that, by the dynamic programming principle, $\xi_i(\tau_j) \in \operatorname{dom} v(\tau_j, .)$ for all $j \in \mathbb{N}$. We now need the following lemma.

Lemma 4.3.6 *For any* $0 < \tau_0 < \tau_1$ *and any pair* (ξ, θ) *solution of*

$$\begin{cases} \xi'(s) = f(s, \xi(s), \theta(s)), \quad \theta(s) \in \mathbb{B}^{n+1} & \text{a.e. } s \in [\tau_0, \tau_1], \\ \xi([\tau_0, \tau_1]) \subset \operatorname{int}\Omega, \\ (\tau_1, \xi(\tau_1)) \in \operatorname{dom} u, \end{cases} \tag{4.41}$$

we have

$$(\xi(t), u(\tau_1, \xi(\tau_1)) + \int_t^{\tau_1} \ell(s, \xi(s), \theta(s))ds) \in \text{epi}\, u(t, .) \quad \forall t \in [\tau_0, \tau_1].$$

Proof Since u is an epigraphical solution, there exists a set $C \subset \mathbb{R}_0^+$ with $\mu_{\mathcal{L}}(C) = 0$ such that for all $(t, x) \in \text{dom}\, u \cap ((\mathbb{R}_0^+ \backslash C) \times \text{int}\, \Omega)$

$$\begin{aligned} & -r + \sup\left\{\langle f(t, x, \theta), -p\rangle + q\ell(t, x, \theta) : \theta \in \mathbb{B}^{n+1}\right\} = 0 \\ & \forall\, (r, p, q) \in T_{\text{epi}\, u}(t, x, u(t, x))^{-}. \end{aligned}$$

Notice that, by the Separation Theorem, this leads us to say that

$$\{-1\} \times (-\Phi(t, x, y)) \subset \text{cl co}\, T_{\text{epi}\, u}(t, x, y)$$

for all $y \geq u(t, x)$ and all $(t, x) \in ((]0, \infty[\backslash C) \times \text{int}\, \Omega) \cap \text{dom}\, u$. Let $0 < \tau_0 < \tau_1$. Thus

$$(1, \bar{f}(t, x, \theta), \bar{\ell}(t, x, \theta)) \in \text{cl co}\, T_{\text{graph}\, E}(t, x, y) \tag{4.42}$$

for a.e. $t \in [0, \tau_1 - \tau_0]$, any $(x, y) \in E(t) \cap (\text{int}\, \Omega \times \mathbb{R})$, and any $\theta \in \mathbb{B}^{n+1}$, where

$$\begin{aligned} \bar{f}(t, x, \theta) &:= -f(\tau_1 - t, x, \theta), \\ \bar{\ell}(t, x, u) &:= \ell(\tau_1 - t, x, \theta), \\ E(t) &:= \text{epi}\, u(\tau_1 - t, \cdot). \end{aligned}$$

Consider a trajectory-control pair $(\xi(.), \theta(.))$ solving (4.41), with $\xi([\tau_0, \tau_1]) \subset \text{int}\, \Omega$ and $(\tau_1, \xi(\tau_1)) \in \text{dom}\, u$. Denote by z the solution of

$$\begin{cases} z'(t) = -\ell(t, \xi(t), \theta(t)) \quad \text{a.e. } t \in [\tau_0, \tau_1], \\ z(\tau_1) = u(\tau_1, \xi(\tau_1)). \end{cases}$$

Put $\bar{\theta}(.) = \theta(\tau_1 - \cdot)$, $\bar{\xi}(.) = \xi(\tau_1 - \cdot)$, and $\bar{z}(.) = z(\tau_1 - \cdot)$ the unique solutions of

$$\begin{cases} \bar{\xi}'(t) = \bar{f}\big(t, \bar{\xi}(t), \bar{\theta}(t)\big), \; \bar{z}'(t) = \bar{\ell}\big(t, \bar{\xi}(t), \bar{\theta}(t)\big) \quad \text{a.e. } t \in [0, \tau_1 - \tau_0], \\ \bar{\xi}(0) = \xi(\tau_1), \quad \bar{z}(0) = u\big(\tau_1, \xi(\tau_1)\big). \end{cases}$$

Applying Theorem 2.2.4 with Φ given by the single-valued map

$$(t, x) \rightsquigarrow \left\{(\bar{f}(t, \bar{\xi}(t), \bar{\theta}(t)), \bar{\ell}(t, \bar{\xi}(t), \bar{\theta}(t)))\right\},$$

the conclusion follows. □

Hence, from the above lemma, we have

$$\int_{t_0}^{\tau_j} \ell(s, \xi_i(s), \theta_i(s))\, ds + u(\tau_j, \xi_i(\tau_j)) \geq u(t_0, \xi_i(t_0)) \quad \forall j \in \mathbb{N}.$$

Recalling (4.32),

$$\int_{t_0}^{\tau_j} \ell(s, \xi_i(s), \theta_i(s))\, ds + \varepsilon \geq u(t_0, \xi_i(t_0)) \quad \forall j \in \mathbb{N},$$

and taking the limit as $j \to +\infty$ we get $\int_{t_0}^{s_i} \ell(s, \xi_i(s), \theta_i(s))\, ds + \varepsilon \geq u(t_0, \xi_i(t_0))$. Passing now to the lower limit as $i \to +\infty$, using (4.39), (4.40), and the lower semicontinuity of u, we have $\int_{t_0}^{+\infty} \ell(s, \bar{\xi}(s), \bar{\theta}(s))\, ds + \varepsilon \geq u(t_0, x_0)$, i.e., $v(t_0, x_0) + \varepsilon \geq u(t_0, x_0)$. Since ε is arbitrary, we conclude

$$v(t_0, x_0) \geq u(t_0, x_0).$$

(C): From parts (A) and (B) we have that $u \leq v \leq w$ on $]0, +\infty[\times\Omega$, that in turn imply $u = w$ on $]0, +\infty[\times\Omega$. Finally, since $t \rightsquigarrow \operatorname{epi} w(t, .)$ is continuous, w is lower semicontinuous, and applying Lemma 1.3.5, we have $\liminf_{s\to 0+,\, y\to_\Omega x} w(s, y) = w(0, x)$ for all $x \in \Omega$. So,

$$w(0, x_0) = \liminf_{\substack{s\to 0+\\ y\to_\Omega x_0}} w(s, y) = \liminf_{\substack{s\to 0+\\ y\to_\Omega x_0}} u(s, y) = u(0, x_0)$$

for any $x_0 \in \Omega$. □

Now, to prove Corollary 4.3.5 we need the following result.

Lemma 4.3.7 *Let* $V : [0, +\infty[\times\Omega \to] - \infty, +\infty]$ *be such that* $t \rightsquigarrow \operatorname{epi} V(t, .)$ *is locally absolutely continuous. Then:*

(i) The following two statements are equivalent:

(i.a) there exists a set $C \subset]0, +\infty[$, *with* $\mu_{\mathscr{L}}(C) = 0$, *such that for all* $(t, x) \in \operatorname{dom} V \cap ((]0, +\infty[\setminus C) \times \Omega)$

$$\exists\, \bar{u} \in \mathbb{B}, \quad D_\uparrow V(t, x)(1, f(t, x, \bar{u})) \leq -\ell(t, x, \bar{u}); \tag{4.43}$$

(i.b) there exists a set $C \subset]0, +\infty[$, *with* $\mu_{\mathscr{L}}(C) = 0$, *such that for all* $(t, x) \in \operatorname{dom} V \cap ((]0, +\infty[\setminus C) \times \Omega)$

$$-p_t + H_{rep}(t, x, -p_x, -q) \geq 0 \quad \forall (p_t, p_x, q) \in T_{\operatorname{epi} V}(t, x, V(t, x))^-;$$

(ii) The following two statements are equivalent:

(ii.a) there exists a set $C \subset]0, +\infty[$, *with* $\mu_{\mathscr{L}}(C) = 0$, *such that for all* $(t, x) \in \operatorname{dom} W \cap ((]0, +\infty[\setminus C) \times \operatorname{int} \Omega)$

$$\forall u \in \mathbb{B}, \quad D_\uparrow V(t, x)(-1, -f(t, x, u)) \leq \ell(t, x, u); \tag{4.44}$$

(ii.b) *there exists a set* $C \subset]0, +\infty[$, *with* $\mu_{\mathscr{L}}(C) = 0$, *such that for all* $(t, x) \in \operatorname{dom} W \cap ((]0, +\infty[\backslash C) \times \operatorname{int} \Omega)$

$$-p_t + H_{rep}(t, x, -p_x, -q) \leq 0 \quad \forall (p_t, p_x, q) \in T_{\operatorname{epi} V}(t, x, V(t, x))^-.$$

Proof See Sect. A.3.3. □

Proof of Corollary 4.3.5 In view of Theorem 4.3.4, the conclusion is a direct consequence of Proposition 4.2.4 and Lemma 4.3.7. □

4.3.2 Connections with Viscosity Solutions

The theory of viscosity solutions provides a framework for analyzing admissible solutions to nonlinear partial differential equations, particularly when classical solutions may not exist due to lack of smoothness. In the context of HJB equations arising from optimal control problems, establishing connections between weak solutions and viscosity solutions is crucial for understanding the relationship between different solution concepts. This section explores these connections and demonstrates how our weak solution framework relates to the well-established viscosity theory.

Proposition 4.3.8 *Consider Assumptions 4.2.1, 4.2.2, and* (4.19). *Let* $V : \mathbb{R}_0^+ \times \Omega \to]-\infty, +\infty[$ *be a locally Lipschitz continuous function satisfying the vanishing condition at infinity*

$$\lim_{t \to +\infty} \sup_{x \in \operatorname{dom} V(t,.)} |V(t, x)| = 0. \tag{4.45}$$

Then the following statements are equivalent:

(i) $V = v$;
(ii) V *is weak solution of the* HJB *equation* (4.30).

Remark 4.3.9 (i) We would like to underline that the outward pointing condition is helpful to construct feasible trajectories for infinite horizon control problems. More precisely, it provides uniform neighboring feasible trajectories results (cfr. Sect. 2.3), on any compact interval $[0, T] \subset \mathbb{R}_0^+$, for the dynamics $\mathscr{F}(s, x) = -f(T - s, x, \mathbb{B})$. Such results basically says that any absolutely continuous trajectory $\xi(.)$ starting from a point in Ω and solving the differential inclusion $\xi'(.) \in \mathscr{F}(., \xi(.))$ can be approximated by a sequence of trajectories which remain in the interior of the state constraints Ω.

(ii) For existence results of the HJB in (4.30) in the free-constraints case we refer to the literature therein Sect. A.4 in Appendix A. Existence results under state constraints are investigate in Chap. 2 in the case of Lagrangian with discount factor $e^{-\lambda t}$ and a suitable inward pointing condition.

We need first the following result.

Lemma 4.3.10 *Consider the assumptions of Proposition 4.3.8. Let* $V : \mathbb{R}_0^+ \times \Omega \to]-\infty, +\infty]$ *be a lower semicontinuous function, satisfying the vanishing condition at infinity* (4.45)*, such that* $\operatorname{dom} v(t,.) \subset \operatorname{dom} V(t,.) \neq \emptyset$ *for all large* $t > 0$ *and*

$$t \rightsquigarrow \{(x,\lambda) \in \Omega \times \mathbb{R} \mid \lambda \leq V(t,x) < +\infty\} \text{ is l.a.c.}$$

Then the following statements are equivalent:

(i) $V = v$;
(ii) $t \rightsquigarrow \operatorname{epi} V(t,.)$ *is l.a.c. and there exists* $C \subset]0,+\infty[$*, with* $\mu_{\mathscr{L}}(]0,+\infty[\setminus C) = 0$*, such that:*

(ii.a) $-p_t + \sup_{u \in \mathbb{B} \subset \mathbb{R}^{n+1}} \{\langle f(t,x,u), -p_x\rangle + q\ell(t,x,u)\} \geq 0$
$\forall (p_t, p_x, q) \in T_{\operatorname{epi} V}(t,x,V(t,x))^-$, $\forall (t,x) \in \operatorname{dom} V \cap (C \times \Omega)$;
(ii.b) $-p_t + \sup_{u \in \mathbb{B} \subset \mathbb{R}^{n+1}} \{\langle f(t,x,u), -p_x\rangle + q\ell(t,x,u)\} \leq 0$
$\forall (p_t, p_x, q) \in T_{\operatorname{hypo} V}(t,x,V(t,x))^+$, $\forall (t,x) \in \operatorname{dom} V \cap (C \times \operatorname{int}\Omega)$.

Proof Notice that, by the definition of locally absolutely continuous set-valued map, the hypograph of $V(t,.)$ restricted to $\operatorname{dom} V(t,.)$ is closed. To show the equivalence between statements (i) and (ii), we use the following claim: for any $(t,x) \in \mathbb{R}_0^+ \times \mathbb{R}^n$ with $\mathscr{U}_\Omega(t,x) \neq \emptyset$, $v(t,x)$ is equal to the following infimum

$$\text{(CV)} \quad \begin{cases} \inf \int_t^{+\infty} H^*(s,\xi(s),\xi'(s))\,ds \text{ over all } \xi \in W^{1,1}_{\mathrm{loc}}(\mathbb{R}_t^+;\mathbb{R}^n) \\ \text{such that } \xi(t) = x \text{ and } \xi(.) \subset \Omega. \end{cases}$$

Indeed, let $(t,x) \in \mathbb{R}_0^+ \times \mathbb{R}^n$ such that $\mathscr{U}_\Omega(t,x) \neq \emptyset$ and denote by $\alpha(t,x) \in [-\infty,+\infty]$ the infimum in (CV) above. From (4.19) we have that $\alpha(t,x) \neq -\infty$. If $\alpha(t,x) = +\infty$ then $\alpha(t,x) \geq v(t,x)$. Assume $\alpha(t,x) \in \mathbb{R}$. Fix $\varepsilon > 0$ and consider $\xi \in W^{1,1}_{\mathrm{loc}}(\mathbb{R}_t^+;\mathbb{R}^n)$ with $\xi(t) = x$ and $\xi(.) \subset \Omega$ satisfying $\int_t^{+\infty} H^*(s,\xi(s),\xi'(s))\,ds < \alpha(t,x) + \varepsilon$. We have that $(\xi'(s), u'(s)) \in \operatorname{graph} H^*(s,\xi(s),.)$ for a.e. $s \geq t$, where we put $u(s) := \int_t^s H^*(\tau,\xi(\tau),\xi'(\tau))\,d\tau$ for all $s \geq t$. Applying now the representation Theorem 4.1.8:(v)' and the Measurable Selection Theorem, we have that there exists a measurable function $w : \mathbb{R}_t^+ \to \mathbb{B}$ such that $(\xi'(s), u'(s)) = (f(s,\xi(s),w(s)), \ell(s,\xi(s),w(s)))$ for a.e. $s \geq t$. We get

$$\int_t^{+\infty} H^*(s,\xi(s),\xi'(s))\,ds = \int_t^{+\infty} u'(s)\,ds = \int_t^{+\infty} \ell(s,\xi(s),w(s))\,ds \geq v(t,x).$$

So, $\alpha(t,x) + \varepsilon > v(t,x)$. Since ε is arbitrary, we deduce that $\alpha(t,x) \geq v(t,x)$. Arguing in analogous way as above and using (4.14), we get $\alpha(t,x) \leq v(t,x)$.

Next, we show the equivalence between the statements (i) and (ii).

Assume (i). Applying Proposition 4.2.4 and the claim, we have that $t \rightsquigarrow \operatorname{epi} v(t,.)$ is l.a.c. for any $x \in \Omega$. Now, from Proposition 4.2.4:(vi) and the claim, we can find a subset $C \subset]0, +\infty[$, with $\mu_{\mathscr{L}}(C) = 0$, such that for any $(t, x) \in (]0, +\infty[\setminus C) \times \operatorname{int}\Omega$ we have $-\ell(t, x, u) \leq D_{\downarrow}v(t, x)(1, f(t, x, u))$ for all $u \in \mathbb{B}$. Hence, from (1.5), we get

$$(1, f(t, x, u), -\ell(t, x, u)) \in T_{\text{hypo}\, v}(t, x, v(t, x))$$

for any $u \in \mathbb{B}$. Then we get

$$\begin{aligned} &- p_t + \sup_{u \in \mathbb{B}} \langle f(t, x, u), -p_x \rangle + q\ell(t, x, u) \leq 0 \\ &\forall (p_t, p_x, q) \in T_{\text{hypo}\, v}(t, x, v(t, x))^+. \end{aligned}$$

Hence, statement (ii.b) holds. Taking into account Lemma 4.2.3, by applying Lemma 4.3.7 and Proposition 4.2.4, we get (ii.a). Thus, (ii) follows.

Next, we show (ii) $\Longrightarrow$ (i). Recalling Lemma 4.3.7 and by inspection of the proof of Theorem 4.3.4, it is sufficient to show the following: there exists $C \subset]0, +\infty[$, with $\mu_{\mathscr{L}}(]0, +\infty[\setminus C) = 0$, such that

$$\begin{aligned} &\forall (t, x) \in \operatorname{dom} V \cap (C \times \operatorname{int}\Omega),\ \forall u \in \mathbb{B}, \\ &D_{\uparrow} V(t, x)(-1, -f(t, x, u)) \leq \ell(t, x, u). \end{aligned} \tag{4.46}$$

From the definition of Φ given in Lemmata 4.2.5, 4.2.3, and the arguments used in the proof of Theorem 4.3.4 applied to the set-valued maps $[0, j] \times \mathbb{R}^n \times \mathbb{R} \ni (s, \xi, \beta) \rightsquigarrow -\Phi(j - s, \xi, \beta) \in \mathbb{R}^n \times \mathbb{R}$ where $j \in \mathbb{N}$, and from the Measurable Selection Theorem, we can find a family of subsets $\tilde{C}_j \subset]0, j[$, with $\mu_{\mathscr{L}}(\tilde{C}_j) = 0$ for all $j \in \mathbb{N}$, such that for any $(t_0, x_0) \in (]0, +\infty[\setminus \cup_{j \in \mathbb{N}} \tilde{C}_j) \times \operatorname{int}\Omega$ and any $u_0 \in \mathbb{B}$, there exists $t_1 \in]0, t_0[$ and a trajectory-control pair $((\xi, \beta), (u, r))(.)$ satisfying

$$\begin{cases} (\xi, \beta)'(t) = (f(t, \xi(t), u(t)), -\ell(t, \xi(t), u(t)) - r(t)) & \text{a.e. } t \in [t_1, t_0], \\ (u, r)(t) \in \mathbb{B} \times [0, c(t)(1 + |\xi(t)|) - \ell(t, \xi(t), u(t))] & \text{a.e. } t \in [t_1, t_0], \\ \xi([t_1, t_0]) \subset \operatorname{int}\Omega, \end{cases}$$

with initial condition and final velocity

$$(\xi, \beta)(t_0) = (x_0, 0), \quad (\xi, \beta)'(t_0) = (f(t_0, x_0, u_0), -\ell(t_0, x_0, u_0)).$$

Now, using the same argument as in Lemma 4.3.6 and since a l.a.c. map is also of LBV, we have the following analogous lemma.

Lemma 4.3.11 *Let* $V : \mathbb{R}_0^+ \times \Omega \to]-\infty, +\infty]$ *be such that*

$$t \rightsquigarrow \{(x, \lambda) \in \Omega \times \mathbb{R} \mid \lambda \leq V(t, x) < +\infty\} \text{ is l.a.c.}$$

If there exists a set $C \subset]0, +\infty[$*, with* $\mu_{\mathscr{L}}(]0, +\infty[\setminus C) = 0$*, such that*

$$\begin{aligned} & -p_t+\sup_{u\in\mathbb{B}}\{\langle f(t,x,u),-p_x\rangle+q\ell(t,x,u)\}\le 0 \\ & \forall\,(p_t,p_x,q)\in T_{\text{hypo}\,V}(t,x,V(t,x))^+,\forall(t,x)\in\text{dom}\,V\cap(C\times\text{int}\,\Omega), \end{aligned} \tag{4.47}$$

then for all $0<\tau_0<\tau_1$ *and any feasible trajectory-control pair* $(\xi(.),u(.))$ *on* $I=[\tau_0,\tau_1]$, *with* $\xi([\tau_0,\tau_1])\subset\text{int}\,\Omega$ *and* $(\tau_0,\xi(\tau_0))\in\text{dom}\,V$, *we have*

$$(\xi(t),V(\tau_0,\xi(\tau_0))-\int_{\tau_0}^{t}\ell(s,\xi(s),u(s))ds)\in\text{hypo}\,V(t,.)\quad\forall t\in[\tau_0,\tau_1].$$

Hence, applying Lemma 4.3.11 and taking a sequence $s_i\in]t_1,t_0[$ with $s_i\to t_0-$, we get $V(s_i,\xi(s_i))-\int_{s_i}^{t_0}\ell(s,\xi(s),u(s))\,ds\le V(t_0,x(t_0))$ for all $i\in\mathbb{N}$. So,

$$V(s_i,\xi(s_i))-V(t_0,x_0)\le\int_{s_i}^{t_0}\ell(s,\xi(s),u(s))\,ds\le\beta(s_i)\quad\forall i\in\mathbb{N}.$$

Dividing by t_0-s_i and passing to the lower limit as $i\to+\infty$, we get (4.46) with $C=]0,+\infty[\setminus\cup_{j\in\mathbb{N}}\tilde{C}_j$, and the proof is complete. □

Proof of Proposition 4.3.8 Let $V:\mathbb{R}_0^+\times\Omega\to\mathbb{R}$ be a locally Lipschitz continuous function and $(t,x)\in\mathbb{R}_0^+\times\Omega$. Notice that, from the locally Lipschitz continuity of V, the following set-valued maps $t\rightsquigarrow\text{epi}\,V(t,.)$ and $t\rightsquigarrow\text{hypo}\,V(t,.)$ are locally absolutely continuous, Since $\partial_-V(t,x)$ and $\partial_+V(t,x)$ are nonempty closed sets, it is straightforward to see that $\cup_{\lambda\ge0}\lambda(\partial_-V(t,x),-1)$ and $\cup_{\lambda\ge0}\lambda(\partial_+V(t,x),-1)$ are closed too. We show that:

$$\cup_{\lambda\ge0}\,\lambda(\partial_-V(t,x),-1)=T_{\text{epi}\,V}(t,x,V(t,x))^-. \tag{4.48}$$

First, we need the following claim. Let φ an extended real-valued function on $\mathbb{R}^m$, l.s.c., proper, and $x\in\text{dom}\,\varphi$, then the following statements are equivalent:

(a) $p\in\partial_-\varphi(x)$;
(b) $(p,-1)\in\widehat{N}_{\text{epi}\,\varphi}(x,\varphi(x))$;
(c) $\langle p,u\rangle\le D_\uparrow\varphi(x)(u)$ for all $u\in\mathbb{R}^m$.

The equivalence between (a) and (b) follows immediately from (1.5) and the definition of regular normal cone. To show that (a) is equivalent to (c): let $p\in\partial_-\varphi(x)$ and sequences $u_k\to u$ and $h_k\to0+$ such that

$$\frac{\varphi(x+h_ku_k)-\varphi(x)}{h_k}\to D_\uparrow\varphi(x)(u).$$

So,

$$\begin{aligned} D_\uparrow\varphi(x)(u)-\langle p,u\rangle &= \lim_{k\to+\infty}\frac{\varphi(x+h_ku_k)-\varphi(x)-h_k\langle p,u_k\rangle}{h_k} \\ &\ge \liminf_{y\to x}\frac{\varphi(y)-\varphi(x)-\langle p,y-x\rangle}{|y-x|}\ge 0. \end{aligned}$$

On the other hand, suppose (c). Consider $y_k \to x$ such that

$$\frac{\varphi(y_k) - \varphi(x) - \langle p, y_k - x \rangle}{|y_k - x|} \to \liminf_{y \to x} \frac{\varphi(y) - \varphi(x) - \langle p, y - x \rangle}{|y - x|}.$$

Passing to a subsequence e keeping the same notation, from Bolzano-Weierstrass Theorem we conclude that $\frac{y_k - x}{|y_k - x|} \to \bar{v} \in \mathbb{S}$. Thus,

$$\begin{aligned}\liminf_{y \to x} \frac{\varphi(y) - \varphi(x) - \langle p, y - x \rangle}{|y - x|} &= \lim_{k \to +\infty} \frac{\varphi(y_k) - \varphi(x)}{|y_k - x|} - \frac{\langle p, y_k - x \rangle}{|y_k - x|} \\ &\geq D_{\uparrow}\varphi(x)(\bar{v}) - \langle p, \bar{v} \rangle \geq 0.\end{aligned}$$

Hence, the claim is proved.

Now, from the claim and Lemma 1.1.2:*(v)* it follows that

$$(\zeta_t, \zeta_x) \in \partial_- V(t, x) \Longleftrightarrow (\zeta_t, \zeta_x, -1) \in T_{\text{epi}\, V}(t, x, V(t, x))^-. \tag{4.49}$$

So,

$$\cup_{\lambda \geq 0}\, \lambda(\partial_- V(t, x), -1) \subset T_{\text{epi}\, V}(t, x, V(t, x))^-.$$

On the other hand, let

$$(\zeta_t, \zeta_x, q) \in T_{\text{epi}\, V}(t, x, V(t, x))^-.$$

Since $(0, 0, \delta) \in T_{\text{epi}\, V}(t, x, V(t, x))$ for all $\delta \geq 0$, we have $q \leq 0$. If $q < 0$, $(\zeta_t/|q|, \zeta_x/|q|, -1) \in T_{\text{epi}\, V}(t, x, V(t, x))^-$ and, applying (4.49), $(\zeta_t/|q|, \zeta_x/|q|) \in \partial_- V(t, x)$. So, $(\zeta_t, \zeta_x, q) \in \cup_{\lambda \geq 0} \lambda(\partial_- V(t, x), -1))$. If $q = 0$, consider $(\bar{\zeta}_t, \bar{\zeta}_x) \in \partial_- V(t, x)$. Then $(\bar{\zeta}_t, \bar{\zeta}_x, -1) \in T_{\text{epi}\, V}(t, x, V(t, x))^-$, and, from the convexity of the polar cone, $(r\bar{\zeta}_t + (1 - r)\zeta_t, r\bar{\zeta}_x + (1 - r)\zeta_x, -r) \in T_{\text{epi}\, V}(t, x, V(t, x))^-$ for all $0 < r < 1$. Arguing as above, we conclude that

$$(r\bar{\zeta}_t + (1 - r)\zeta_t, r\bar{\zeta}_x + (1 - r)\zeta_x, -r) \in \cup_{\lambda \geq 0} \lambda(\partial_+ V(t, x), -1) \quad \forall r \in]0, 1[,$$

and the claim (4.48) follows.

Using the same argument as above, we have also

$$\cup_{\lambda \geq 0}\, \lambda(\partial_+ V(t, x), -1) = T_{\text{hypo}\, V}(t, x, V(t, x))^+. \tag{4.50}$$

Finally, from (4.48), (4.50), Lemma 4.3.10, and the representation Theorem 4.1.8, the conclusion follows. □

Appendix

A.1 Necessary Conditions on Finite Horizon

We recall known necessary conditions in presence of functional constraints for non-smooth optimal control problems on finite horizon.

Theorem A.1.1 *Let* $\bar{\mathrm{x}} \in W^{1,1}([S,T];\mathbb{R}^k)$ *be a local minimizer for the problem,*

$$\begin{cases} \textit{minimize } \phi(\mathrm{x}(S),\mathrm{x}(T)) \\ \textit{over all } \mathrm{x} \in W^{1,1}([S,T];\mathbb{R}^k) \textit{ satisfying} \\ \mathrm{x}'(t) \in \mathcal{F}(t,\mathrm{x}(t)) \textit{ a.e. } t \in [S,T], \\ (\mathrm{x}(S),\mathrm{x}(T)) \in C \textit{ closed subset}, \\ g(\mathrm{x}(t)) \le 0 \ \textit{ for all } t \in [S,T]. \end{cases}$$

Assume that there exists $\varepsilon > 0$ *such that:*

– ϕ *is Lipschitz continuous on a neighborhood of* $(\bar{\mathrm{x}}(S),\bar{\mathrm{x}}(T))$*;*
– $\mathcal{F}(t,\mathrm{x}) \neq \emptyset$ *for all* (t,x)*,* $\mathcal{F}(.,.)$ *is Lebesgue-Borel measurable, graph* $\mathcal{F}(t,.)$ *is closed for each* $t \in [S,T]$*, and there exist* $k \in L^1([S,T];\mathbb{R}_0^+)$ *such that, for a.e.* $t \in [S,T]$*,*

$$\mathcal{F}(t,\tilde{\mathrm{x}}) \subset \mathcal{F}(t,\mathrm{x}) + k(t)\,|\tilde{x}-x|\,\mathbb{B}$$

for all $x, \tilde{\mathrm{x}} \in \bar{\mathrm{x}}(t)+\varepsilon\mathbb{B}$*;*
– *there is a constant* $c > 0$ *such that*

$$|g(\mathrm{x}) - g(\tilde{\mathrm{x}})| \le c\,|\mathrm{x}-\tilde{\mathrm{x}}|$$

for all $\mathrm{x}, \tilde{\mathrm{x}} \in \{\mathrm{x} \in \mathbb{R}^k \mid \mathrm{x} \in \bar{\mathrm{x}}(t)+\varepsilon\mathbb{B}, t \in [S,T]\}$.

Then there exist an arc $p \in W^{1,1}\left([S,T];\mathbb{R}^k\right)$*, a number* $\lambda \ge 0$*, a nonnegative Borel measure* μ *on* $[S,T]$ *with* $\operatorname{supp}\mu \subset \{t \mid g(\bar{\mathrm{x}}(t)) = 0\}$*, and a* μ*-integrable function* $\gamma(.)$*, such that:*

V. Basco, *Nonsmooth Constrained Optimal Control on Infinite Horizon*, SpringerBriefs in Mathematics, https://doi.org/10.1007/978-3-032-08938-0

(i) $\lambda + \|p\|_{\infty,[S,T]} + \int_{[S,T]} \mu(dt) = 1$;
(ii) $p'(t) \in \text{co}\left\{\eta \mid (\eta, p(t) + \int_{[S,t[} \gamma(s)\mu(ds)) \in N_{\text{graph}\,\mathcal{F}(t,.)}(\bar{\mathrm{x}}(t), \bar{\mathrm{x}}'(t))\right\}$ *a.e.* t;
(iii) $(p(S), -(p(T) + \int_{[S,T]} \gamma(s)\mu(ds))) \in \lambda\partial\phi(\bar{\mathrm{x}}(S), \bar{\mathrm{x}}(T)) + N_C(\bar{\mathrm{x}}(S), \bar{\mathrm{x}}(T))$;
(iv) $\langle\, p(t) + \int_{[S,t[} \gamma(s)\mu(ds), \bar{\mathrm{x}}'(t)\rangle \geq \langle\, p(t) + \int_{[S,t[} \gamma(s)\mu(ds), v\rangle$ *for all* $v \in \mathcal{F}(t, \bar{\mathrm{x}}(t))$ *a.e.* t;
(v) $\gamma(t) \in \partial^{\sharp}_{\mathrm{x}} g(t, \bar{\mathrm{x}}(t))$ μ*-a.e. where*

$$\partial^{\sharp}_{\mathrm{x}} g(t, \mathrm{x}) := \text{cl co}\{\xi \mid \exists \xi_i \to \xi, \exists\, \mathrm{x}_i \to \mathrm{x},\ \textit{such that}\ \xi_i = \nabla g(t, \mathrm{x}_i)\ \textit{and}\ g(t, \mathrm{x}_i) > 0\ \textit{for each}\ i\}.$$

Proof See [23, Chap. 10.3] or [13, Chap. 5.2]). □

A.2 A Result on Upper Semicontinuous Set-Valued Maps

The following result states that upper semicontinuous set-valued maps Φ with closed convex values are σ-selectionable, i.e. there exists a sequence of decreasing set-valued maps Φ_n with compact values and closed graph such that Φ_n has a continuous selection (for any n) and $\Phi = \cap_n \Phi_n$.

Lemma A.2.1 *Let X be a metric space, Y be a Banach space, and $\Phi : X \rightsquigarrow Y$ be a set-valued map upper semicontinuous with convex closed values. Then there exists a sequence of upper semicontinuous maps $\Phi_n : X \rightsquigarrow \text{cl co}(\Phi(X))$ which approximate Φ in the sense that for all $x \in X$:*

- *$\forall n \in \mathbb{N}$, $\Phi(x) \subset \Phi_{n+1}(x) \subset \Phi_n(x)$;*
- *$\forall \varepsilon > 0$, $\exists N(\varepsilon, x)$ such that $\forall n \geq N(\varepsilon, x)$, $\Phi_n(x) \subset \Phi(x) + \varepsilon\mathbb{B}$.*

The maps Φ_n can be written in the following form:

$$\forall x \in X, \quad \Phi_n(x) := \sum_i g_i^n(x) C_i^n$$

where the subsets $\{C_i^n\}_i$ are closed and convex and where the functions $\{g_i^n\}_i$ form a locally Lipschitzean locally finite partition of unity.

Proof See [1, Chap. 1, Sect. 13, Theorem 1]. □

A.3 Supporting Proofs

The following lemmata are needed in the subsequent proofs.

Lemma A.3.1 *Consider two complete separable metric spaces X, Y, and measurable set-valued maps $\Phi : \Theta \rightsquigarrow X$, $\Psi : \Theta \rightsquigarrow Y$, with closed images. Let $\psi : \Theta \times X \to Y$ be a Carathéodory map. Then the set-valued map Ξ defined by*

$$\Xi(\theta) := \{x \in \Phi(\theta) \mid \psi(\theta, x) \in \Psi(\theta)\}$$

is measurable. Consequently, if

$$\forall \theta \in \Theta, \quad \psi(\theta, \Phi(\theta)) \cap \Psi(\theta) \neq \emptyset$$

then there exists a measurable selection ϕ of Φ such that $\psi(\theta, \phi(\theta)) \in \Psi(\theta)$ for every $\theta \in \Theta$.

Proof See [12, Theorem III.38, p. 85]. □

Lemma A.3.2 *For every sequence of measurable real-valued functions $\phi_n : \Theta \to \mathbb{R}$, the extended function*

$$\theta \mapsto \phi(\theta) := \inf_{n \in \mathbb{N}^+} \phi_n(\theta) \in [-\infty, +\infty[$$

has a measurable domain of definition and is measurable on it. A similar statement holds true for $\sup_{n \in \mathbb{N}^+} \phi_n$.

Proof See [16, Proposition 2.7, p. 45]. □

A.3.1 Proofs of Chap. 2

Proof of Lemma 2.2.5 Let us denote by with $\Phi_\sharp$ the set-valued map as in Lemma 1.3.1. We have

$$\mathrm{Ker}(\Phi, s, x) = \mathrm{Ker}(\Phi_\sharp, s, x).$$

Fix $\delta > 0$. Let $C_\delta \subset I$ be a closed subset for which $\mu_{\mathscr{L}}(I \backslash C_\delta) < \delta$, $\Phi_\sharp$ is upper semicontinuous on $C_\delta \times \mathbb{R}^d$, and

$$\Phi_\sharp(t, x) \subset \Phi(t, x) \quad \forall (t, x) \in C_\delta \times \mathbb{R}^d. \tag{A.1}$$

We can assume that $\mu_{\mathscr{L}}(C_\delta) > 0$. Define the nonempty sets[3]

[3] Let $f \in L^1_{\mathrm{loc}}(\mathbb{R}^n; \mathbb{R})$. A point x is called a *Lebesgue point* for f if

$$\widetilde{W}_\delta := \{p \in I \mid p \text{ is a Lebesgue point for } q(.)\chi_{I\setminus C_\delta}(.)\},$$
$$\widetilde{G}_\delta := \{p \in I \mid p \text{ is a Lebesgue point for } \chi_{C_\delta}(.)\}.$$

Denote $W_\delta := C_\delta \cap \widetilde{W}_\delta$ and $G_\delta := W_\delta \cap \widetilde{G}_\delta$. We have $\mu_{\mathscr{L}}(I\setminus\widetilde{W}_\delta) = \mu_{\mathscr{L}}(I\setminus\widetilde{G}_\delta) = 0$. Moreover, $\mu_{\mathscr{L}}(C_\delta \cap \widetilde{W}_\delta) = \mu_{\mathscr{L}}(C_\delta) - \mu_{\mathscr{L}}(C_\delta\setminus\widetilde{W}_\delta)$ and $\mu_{\mathscr{L}}(C_\delta\setminus\widetilde{W}_\delta) \leq \mu_{\mathscr{L}}(I\setminus\widetilde{W}_\delta) = 0$ yield $\mu_{\mathscr{L}}(C_\delta \cap \widetilde{W}_\delta) = \mu_{\mathscr{L}}(C_\delta)$. Arguing in the same way, we deduce also that $\mu_{\mathscr{L}}(W_\delta \cap \widetilde{G}_\delta) = \mu_{\mathscr{L}}(W_\delta)$. So, $\mu_{\mathscr{L}}(G_\delta) = \mu_{\mathscr{L}}(W_\delta) = \mu_{\mathscr{L}}(C_\delta) > 0$ and in particular $G_\delta \neq \emptyset$. We fix $\tau \in G_\delta$ and $x_\tau \in \mathbb{R}^d$. Notice that (cfr. [12, Chap. VI, Sect. 4] for the existence and compactness of continuous solutions under our assumptions) we can find $M > 0$ such that for any $x \in \text{Ker}\,(\Phi, \tau, x_\tau)$, $|x(t)| \leq M$ for all $t \in I$.

Next, we consider only $t < \tau$, since the case $t > \tau$ follows by the same arguments. We have

$$\frac{1}{t-\tau}\int_{[\tau,t]} \chi_{I\setminus C_\delta}(s)q(s)ds \to 0, \quad \frac{\mu_{\mathscr{L}}([\tau,t]\cap C_\delta)}{t-\tau} \to 1, \quad \text{as } t \to \tau+. \tag{A.2}$$

Let $\varepsilon > 0$. Thanks to Lemma 1.3.1, there exists $\eta_1 > 0$ such that for any $x \in \text{Ker}\,(\Phi, \tau, x_\tau)$

$$\Phi_\sharp(s, x(s)) \subset \Phi_\sharp(\tau, x_\tau) + \frac{\varepsilon}{4}\mathbb{B} \quad \forall s \in C_\delta \cap]\tau, \tau + \eta_1[. \tag{A.3}$$

Recalling (A.2), there exists $\eta_2 > 0$ such that if $t \in]\tau, \tau + \eta_2[$ then

$$\frac{1+M}{t-\tau}\int_{[\tau,t]\setminus C_\delta} q(s)ds < \frac{\varepsilon}{4}, \tag{A.4}$$

and there exists $\eta_3 > 0$ such that if $t \in]\tau, \tau + \eta_3[$ then

$$\begin{cases} \frac{\mu_{\mathscr{L}}([\tau,t]\cap C_\delta)}{t-\tau}\Phi(\tau, x_\tau) \subset \Phi(\tau, x_\tau) + \frac{\varepsilon}{4}\mathbb{B}, \\ \frac{\mu_{\mathscr{L}}([\tau,t]\cap C_\delta)}{t-\tau}\frac{\varepsilon}{4} \leq \frac{\varepsilon}{2}. \end{cases} \tag{A.5}$$

Let now $t \in]\tau, \tau + \eta[$ with $\eta := \min\{\eta_1, \eta_2, \eta_3\}$. Then for any $x \in \text{Ker}\,(\Phi, \tau, x_\tau)$ we get

$$\lim_{\substack{\mu_{\mathscr{L}}(Q)\to 0 \\ x\in Q}} \frac{1}{\mu_{\mathscr{L}}(Q)}\int_Q |f(y) - f(x)|dy = 0$$

where Q stands for n-dimensional cubes. It is well known that x is a Lebesgue point for f for a.e. $x \in \mathbb{R}^n$ and holds $\lim_{\substack{\mu_{\mathscr{L}}(Q)\to 0 \\ x\in Q}} \frac{1}{\mu_{\mathscr{L}}(Q)}\int_Q f(y)dy = f(x)$.

$$
\begin{aligned}
\frac{x(t)-x_\tau}{t-\tau} &\in \frac{1}{t-\tau}\int_{[\tau,t]\cap C_\delta}\Phi_\sharp(s,x(s))ds + \frac{1}{t-\tau}\int_{[\tau,t]\setminus C_\delta}\Phi_\sharp(s,x(s))ds\\
&\underset{(A.1),(A.3)}{\subset} \frac{\mu_{\mathscr{L}}([\tau,t]\cap C_\delta)}{t-\tau}\left(\Phi(\tau,x_\tau)+\frac{\varepsilon}{4}\mathbb{B}\right)+\frac{1+M}{t-\tau}\int_{[\tau,t]\setminus C_\delta} q(s)ds\,\mathbb{B}\\
&\underset{(A.4),(A.5)}{\subset} \Phi(\tau,x_\tau)+\frac{\varepsilon}{2}\mathbb{B}+\frac{\varepsilon}{4}\mathbb{B}+\frac{\varepsilon}{4}\mathbb{B}=\Phi(\tau,x_\tau)+\varepsilon\mathbb{B}.
\end{aligned}
$$

Then the first statement follows keeping $\mathscr{B} = \cup_{i\in\mathbb{N}^+} G_{1/i}$.

Next, we show the second statement. Let $\phi(.,.,.)$ be a Charathéodory parametric representation of Φ given by [15] (see also the proof of representation utilizing the Steiner projection as in Theorem 4.1.1). Such representation satisfies for all t

$$
\begin{aligned}
&|\phi(t,x,\omega)-\phi(t,y,\lambda)|\\
&\le C(\|\Phi(t,x)\||\omega-\lambda| + d_{\mathscr{H}}(\Phi(t,x),\Phi(t,y)) + \|\Phi(t,x)\|\omega - \|\Phi(t,y)\|\lambda),
\end{aligned}
$$

for a suitable constant $C > 0$. Since, by assumption, $\|\Phi(t,x)\| \le q(t) < +\infty$ and $\Phi(t,.)$ is continuous, then $x \rightsquigarrow \|\Phi(t,x)\|$ is continuous. Hence, $\phi(t,.,.)$ is continuous for every t.

Now, let any $\tau \in G_\delta$ as above, $x_\tau \in \mathbb{R}^d$, $\overline{\phi} \in \Phi(\tau,x_\tau)$, and $\overline{\omega} \in \mathbb{B}$ such that $\phi(\tau,x_\tau,\overline{\omega}) = \overline{\phi}$. From Lemma 1.3.1 applied to the single valued map $(t,x,\omega) \rightsquigarrow \{\phi(t,x,\omega)\}$, there exists a function $\phi_\sharp$, continuous on $C_\delta \times \mathbb{R}^n \times \mathbb{B}$, with $\phi_\sharp = \phi$ on $C_\delta \times \mathbb{R}^n \times \mathbb{B}$. Let any solution of $x'(s) = \phi(s,x(s),\overline{\omega})$ a.e. s in a neighborhood of τ satisfying $x(\tau) = x_\tau$. We have

$$
\begin{aligned}
&\frac{x(t)-x_\tau}{t-\tau}\\
&= \frac{1}{t-\tau}\int_{[\tau,t]\cap C_\delta}\phi_\sharp(s,x(s),\overline{\omega}) - \phi_\sharp(t,x_\tau,\overline{\omega})ds\\
&\quad + \frac{1}{t-\tau}\int_{[\tau,t]\cap C_\delta}\phi_\sharp(t,x_\tau,\overline{\omega})ds\\
&\quad + \frac{1}{t-\tau}\int_{[\tau,t]\setminus C_\delta}\phi_\sharp(s,x(s),\overline{\omega})ds.
\end{aligned}
$$

Then, using the same arguments as in the first part, we get the conclusion. □

A.3.2 Proofs of Chap. 3

Proof of Lemma 3.2.4 Consider a dense family of measurable selections $(\psi_n)_{n\in\mathbb{N}}$ of the map Φ (cfr. [12, Theorem III.30, p. 80]) i.e. $\mathrm{cl}(\{\psi_n(\theta) \mid n \in \mathbb{N}\}) = \Phi(\theta)$ for every $\theta \in \Theta$. Since the maps $\phi(\theta,.)$ are continuous, we get that

$$
\forall \theta \in \Theta, \quad \alpha(\theta) = \inf_{n\in\mathbb{N}} \phi(\theta,\psi_n(\theta)).
$$

Now, we claim that the maps $\theta \mapsto \phi(\theta, \psi_n(\theta))$ are measurable for every $n \in \mathbb{N}$. Indeed, for any $n \in \mathbb{N}$, since $\psi_n(.)$ is measurable, there exists a sequence of simple measurable maps $(\psi_k^{(n)})_k : \Theta \to X$ converging pointwise to $\psi_n(.)$. Then $\theta \mapsto \phi(\theta, \psi_k^{(n)}(\theta))$ is measurable. Since ϕ is continuous with respect to the second variable,

$$\forall n \in \mathbb{N}, \forall \theta \in \Theta, \quad \lim_{k \to +\infty} \phi(\theta, \psi_k^{(n)}(\theta)) = \phi(\theta, \psi_n(\theta)).$$

Hence $\phi(\theta, \psi_n(\theta))$ is the pointwise limit of measurable maps and so, is measurable. Now, from Lemma A.3.2, we get that $\alpha(.)$ is also measurable. Applying Lemma A.3.1 we get that the marginal map Ξ is measurable, since

$$\Xi(\theta) = \{x \in \Phi(\theta) \mid \phi(\theta, x) = \alpha(\theta)\}$$

and because $\alpha(.)$ is measurable. □

Proof of Lemma 3.2.6 For simplicity, we formally consider the interval $[S, T] = [0, 1]$. Extend $q(.)$ to all of $\mathbb{R}$, zero outside the interval $[0, 1]$. Consider $K \geq 0$ such that $\|q\|_{\infty,[0,1]} \leq K$. Take $\varepsilon > 0$. By Lusin's Theorem in the vector case, there exist a function $\tilde{q}$ and a measurable set Ω_ε such that: $\tilde{q}$ is continuous on $[0, 1]$, taking value 0 off $[0, 1]$ and having values bounded by K; $\mu_{\mathcal{L}}(\Omega_\varepsilon) \leq \varepsilon$; $\tilde{q}(s) = q(s)$ for all $s \in [0, 1]\backslash\Omega_\varepsilon$. So, for any $h \in \mathbb{R}$

$$\begin{aligned}\int_0^1 |q(s+h) - q(s)|ds &\leq \int_0^1 |\tilde{q}(s+h) - \tilde{q}(s)|ds \\ &\quad + \int_0^1 (|\tilde{q}(s+h) - q(s+h)| + |\tilde{q}(s) - q(s)|)ds \\ &\leq \int_0^1 |\tilde{q}(s+h) - \tilde{q}(s)|ds + 4K\varepsilon.\end{aligned}$$

By the Dominated Convergence Theorem, the integrals on the right-hand side have limit 0 as $h \to 0$, since the integrals are majored by the constant function $2K$ and since they converge point-wise to zero except possibly at $s \in \{0, 1\}$. We get $\limsup_{h \to 0} \int_0^1 |q(s+h) - q(s)|ds \leq 4K\varepsilon$. Since $\varepsilon > 0$ was arbitrary, it follows that

$$\lim_{h \to 0} \int_0^1 |q(s+h) - q(s)|ds = 0. \tag{A.6}$$

Now define

$$\Delta_N(\tau) := \int_0^1 |q\,(\tau + \psi_N(s - \tau)) - q(s)|\,ds.$$

Then, taking into account the variable changing $\sigma = s - \tau$,

$$\int_0^1 \Delta_N(\tau)d\tau = \int_0^1 \int_0^1 |q(\tau + \psi_N(s-\tau)) - q(s)|\, ds d\tau$$
$$= \int_{-1}^{+1} \left(\int_{0\vee\sigma}^{1\wedge(1+\sigma)} |q(s - \sigma + \psi_N(\sigma)) - q(s)|\, ds \right) d\sigma.$$

Notice that, from (A.6), for any $\sigma \in \mathbb{R}$ we have

$$\lim_{N\to+\infty} \int_{0\vee\sigma}^{1\wedge(1+\sigma)} |q(s - \sigma + \psi_N(\sigma)) - q(s)| ds$$
$$\leq \lim_{N\to+\infty} \int_0^1 |q(s - \sigma + \psi_N(\sigma)) - q(s)|\, ds = 0.$$

Applying again the Dominated Convergence Theorem

$$\lim_{N\to+\infty} \int_0^1 \Delta_N(\tau) = 0.$$

Hence, there exists a subsequence $(N_j)_j$ and a set $\mathscr{S}$ of full measure, such that $\Delta_{N_j}(\tau) \to 0$ for all $\tau \in \mathscr{S}$. Since

$$\sup_{t\in[0,1]} \left| \int_0^t (q(\tau + \psi_{N_j}(s-\tau)) - q(s))ds \right| \leq \Delta_{N_j}(\tau) \quad \forall \tau \in \mathscr{S},$$

the (3.10) follows.

Now, recalling that

$$\int_{[0,1]} dq_{N,\tau}(s) = \sum_{\{j \mid \tau + \frac{j}{N} \in [0,1]\}} q\left(\tau + \frac{j}{N}\right) \cdot \frac{1}{N}, \tag{A.7}$$

from the definitions of $q_{N,\tau}$, $\tilde{q}_{N,\tau}$, for any $t \in \mathbb{R}$, $\tau \in [0, 1]$, and $N \in \mathbb{N}$, we have

$$\left| \int_0^t \tilde{q}_{N,\tau}(s)ds - \int_0^t dq_{N,\tau}(s) \right| \leq \sup_{s\in[0,1]} \frac{q(s)}{N}.$$

Thus, from (3.10) and since q is uniformly bounded, we get (3.11), and the proof is complete. □

Proof of Lemma 3.3.4 First of all, invoching Riesz–Markov–Kakutani Theorem, we regard the nonnegative Borel measures μ as elements of the dual of $C([S, T]; \mathbb{R})$ equipped with the induced total variation norm (i.e. $\int_{[S,T]} \mu(dt)$). Now, since the total variations of μ_i are uniformly bounded (due to the weakly* convergence) and since there exists $M > 0$ with $|\Pi_i(t)| \leq M$ for μ_i-a.e. t and all $i \in \mathbb{N}$, we deduce that the total variation (induced from the Euclidean norm) of the vector valued measures $\Pi_i(t)\mu_i(dt)$ are also uniformly bounded. So, there exists a Borel measure Φ_0 such that the measures $\Phi_i := \Pi_i \mu_i$ weakly* converges to Φ_0. Hence, for any continuous function $f : [S,T] \to \mathbb{R}^n$

$$\begin{aligned}
\left|\int_{[S,T]} f(t)\Phi_0(dt)\right| &= \lim_{i\to+\infty} \left|\int_{[S,T]} f(t)\Pi_i(t)\mu_i(dt)\right| \\
&\leq M \lim_{i\to+\infty} \int_{[S,T]} |f(t)|\mu_i(dt) \\
&\leq M \int_{[S,T]} |f(t)|\mu_0(dt).
\end{aligned}$$

So, Φ_0 is absolutely continuous wrt μ_0 and by Radon-Nikodym Theorem there exists a μ_0-measurable function $\Pi_0 : [S,T] \to \mathbb{R}^n$ such that

$$\int_{\mathscr{E}} \Phi_0(dt) = \int_{\mathscr{E}} \Pi_0(t)\mu_0(dt)$$

for every Borel measurable set $\mathscr{E} \subset [S,T]$. Now, notice that for any $p \in \mathbb{R}^n$ the function $z_p : [S,T] \to \mathbb{R}$

$$z_p(t) := \max\{\langle p, w\rangle \mid w \in A(t)\} \quad \forall t \in [S,T]$$

is upper semicontinuous. Let any $p \in \mathbb{R}^n$ and consider a sequence (see [1, Theorem A6.6]) of continuous functions $z_j^{(p)} : [S,T] \to \mathbb{R}$ such that for all $t \in [S,T]$

$$\begin{cases} z_p(t) \leq z_j^{(p)}(t) \quad \forall j \in \mathbb{N}, \\ z_p(t) = \lim_{j\to+\infty} z_j^{(p)}(t). \end{cases}$$

Fix $j \in \mathbb{N}$. Hence, for all $i \in \mathbb{N}$

$$\begin{aligned}
\langle p, \int_{\mathscr{E}} \Phi_i(dt)\rangle &= \langle p, \int_{\mathscr{E}} \Pi_i(t)\mu_i(dt)\rangle \\
&= \int_{\mathscr{E}} \langle p, \Pi_i(t)\rangle\mu_i(dt) \leq \int_{\mathscr{E}} z_p(t)\mu_i(dt) \leq \int_{E} z_j^{(p)}(t)\mu_i(dt)
\end{aligned}$$

for every Borel measurable set $\mathscr{E}$. Keeping subsequences and using the same notation, passing to the limit as $i \to +\infty$

$$\langle p, \int_{\mathscr{E}} \Phi_0(dt)\rangle = \int_{\mathscr{E}} \langle p, \Pi_0(t)\rangle\mu_0(dt) \leq \int_{\mathscr{E}} z_j^{(p)}(t)\mu_0(dt)$$

for every Borel measurable set $\mathscr{E}$. So, we deduce that $\langle p, \Pi_0(t)\rangle \leq z_j^{(p)}(t)$ for μ_0-a.e. t. Passing to the limit as $j \to +\infty$, we have that

$$\langle p, \Pi_0(t)\rangle \le z_p(t) \text{ for } \mu_0\text{-a.e. } t.$$

We can conclude that there exists a dense subset $Q \subset \mathbb{R}^n$ and a Borel measurable set $\mathscr{C} \subset [S, T]$, with $\mu_0(\mathscr{C}) = 0$, such that

$$\langle p, \Pi_0(t)\rangle \le z_p(t) \quad \forall p \in Q, \forall t \in [S, T]\backslash\mathscr{C}.$$

Let any $p \in \mathbb{R}^n$ and $t \in [S, T]\backslash\mathscr{C}$ and keep $p_j \to_Q p$. We have $\langle p_j, \Pi_0(t)\rangle \le z_{p_j}(t)$ and passing to the limit as $j \to +\infty$ we get $\langle p, \Pi_0(t)\rangle \le z_p(t)$ because of $p \mapsto z_p(t)$ is continuous for every t. So, $\langle p, \Pi_0(t)\rangle \le z_p(t)$ for all $p \in \mathbb{R}^n$ and all $t \in [S, T]\backslash\mathscr{C}$. Since $A(t)$ is nonempty compact convex, by applying the Hahn-Banach Theorem we finally conclude that $\Pi_0(t) \in A(t)$ for μ_0-a.e. t. □

A.3.3 Proofs of Chap. 4

Proof of Lemma 4.1.2 Let $\mathscr{O} \subset X$ be an open set. Then

$$(\theta \rightsquigarrow \text{cl}\,(\bigcup_{n\in\mathbb{N}} \Phi_n(\theta)))^{-1}(\mathscr{O}) = \left\{\theta \in \Theta \mid \bigcup_{n\in\mathbb{N}} (\Phi_n(\theta) \cap \mathscr{O}) \neq \emptyset\right\} = \bigcup_{n\in\mathbb{N}} \Phi_n^{-1}(\mathscr{O}).$$

To prove the second statement observe that $\cap_{n\in\mathbb{N}}\Phi_n(\theta)$ is closed for any $\theta \in \Theta$ and

$$\text{graph}(\theta \rightsquigarrow \bigcap_{n\in\mathbb{N}} \Phi_n(\theta)) = \bigcap_{n\in\mathbb{N}} \text{graph}\ \Phi_n.$$

By the Characterization of Measurability of Set-valued Maps, graph Φ_n belongs to $\mathfrak{A} \otimes \mathfrak{B}$ (here $\mathfrak{B}$ is the Borel σ-algebra of the metric space X) and so does the graph of $\theta \rightsquigarrow \cap_{n\in\mathbb{N}}\Phi_n(\theta)$, which is measurable, thanks to the same result. □

Proof of Lemma 4.1.3 Consider the function $\psi(\theta, x) = \mathfrak{d}(x, \phi(\theta))$. As in the proof of Lemma 3.2.4, we have that $\psi(., x)$ is measurable. Since it is also continuous in x Lemma we may use A.3.1 with $\Phi \equiv X$, $\Psi(\theta) = [0, \rho(\theta)]$. To prove the second and third statements, we apply Lemma 3.2.4 to the function $(\theta, x) \mapsto \mathfrak{d}(x, \phi(\theta))$ □

The following proof is based on [21, Lemma 1, p. 37], in which the author derives sharper estimates (cfr. also [19, 20]).

Proof of Lemma 4.1.4 We first show that: for all convex nonempty K, L

$$d_{\mathscr{H}}(P(0, K), P(0, L)) \le 5d_{\mathscr{H}}(K, L). \tag{A.8}$$

Indeed, let K, L convex nonempty such that $d_{\mathscr{H}}(K, L) < +\infty$ and put $\varepsilon := d_{\mathscr{H}}(K, L)$. If $\varepsilon = 0$, then $K = L$ and the claim follows. Next, assume that $\varepsilon > 0$. Fix $y \in P(0, L)$, $x_1 \in K$ with $|y - x_1| \le \varepsilon$, and $x_2 \in K$ such that $d_K(0) = |x_2|$.

We need to show that:

$$\exists x \in P(0, K) : \quad |x - y| \leq 5\varepsilon. \tag{A.9}$$

Observe first that

$$|y| \leq 2d_L(0) \leq 2d_K(0) + 2\varepsilon. \tag{A.10}$$

If $|x_1| \leq 2d_K(0)$ then set $x := x_1$ and (A.9) holds. Otherwise, if $x_2 = 0$ set $x := 0$. Then (A.10) implies (A.9). It remains to consider the case

$$|x_1| > 2d_K(0) \quad \& \quad x_2 \neq 0.$$

Let us define the continuous function $\varphi : [0, 1] \to \mathbb{R}^k$ by

$$\forall \lambda \in [0, 1], \quad \varphi(\lambda) = |\lambda x_1 + (1 - \lambda)x_2|.$$

Notice that, since the image of φ is a connected and

$$\varphi(0) = d_K(0) \quad \& \quad \varphi(1) > 2d_K(0),$$

then there exists $\bar{\lambda} \in]0, 1[$ satisfying $\varphi(\bar{\lambda}) = 2d_K(0)$. Hence, setting $x := \bar{\lambda}x_1 + (1 - \bar{\lambda})x_2$, it follows that $x \in [x_1, x_2]$ and

$$|x| = 2d_K(0).$$

The set K being convex, $x \in P(0, K)$. Now, by the choice of x_1,

$$|x - x_1| \geq |x - y| - |y - x_1| \geq |x - y| - \varepsilon. \tag{A.11}$$

In the triangle $T(0; x; x_2)$[1] denote by α the angle at x. Inequality $|x_2| < |x|$ yields that $\alpha \in \left[0, \frac{\pi}{2}\right[$ and

$$\sin\alpha \leq \frac{|x_2|}{|x|} = \frac{1}{2}, \quad \cos\alpha \geq \frac{\sqrt{3}}{2}. \tag{A.12}$$

We now distinguish two cases, $\alpha = 0$ and $\alpha > 0$.

Case $\alpha = 0$: We have $|x_1| = |x| + |x - x_1|$ and (A.11) implies

$$|y| \geq |x_1| - |y - x_1| \geq |x| + |x_1 - x| - \varepsilon \geq 2d_K(0) + |x - y| - 2\varepsilon.$$

So, we deduce from (A.10) that

$$|x - y| \leq 4\varepsilon$$

[1] $T(x; y; z)$ denotes the triangle with vertices at x, y, z.

and get (A.9).

Case $\alpha > 0$: Then the angle of $T(0; x; x_1)$ at x is equal to $\pi - \alpha > \frac{\pi}{2}$. Let $z \in]0, x_1[$ be such that $\langle z - x, x \rangle = 0$ (i.e. $z - x$ is orthogonal to x). Then

$$|z| \geq |x| = 2d_K(0). \tag{A.13}$$

In the triangle $T(z; x; x_1)$, the angle at x is equal to $\pi - \alpha - \frac{\pi}{2} = \frac{\pi}{2} - \alpha$. Therefore (A.12) yield

$$|z - x_1| \geq |x - x_1| \sin\left(\frac{\pi}{2} - \alpha\right) = |x - x_1| \cos\alpha \geq |x - x_1| \frac{\sqrt{3}}{2},$$

and from (A.13) follows that

$$|x_1| = |z| + |z - x_1| \geq 2d_K(0) + \frac{\sqrt{3}}{2}|x - x_1|.$$

This and (A.11) imply

$$|y| \geq |x_1| - |y - x_1| \geq 2d_K(0) + \frac{\sqrt{3}}{2}(|x - y| - \varepsilon) - \varepsilon.$$

Therefore, thanks to (A.10), we derive

$$\frac{\sqrt{3}}{2}(|x - y| - \varepsilon) - \varepsilon \leq 2\varepsilon.$$

So, (A.9) holds true. This conclude the proof of (A.8).

Now, let any x, y and notice that

$$d_{\mathscr{H}}(P(x, K), P(x, L)) = d_{\mathscr{H}}(P(0, K - x), P(0, L - x)).$$

Using (A.8), we deduce that

$$\begin{aligned}
&d_{\mathscr{H}}(P(x, K), P(y, L)) \\
&\leq d_{\mathscr{H}}(P(x, K), P(x, L)) + d_{\mathscr{H}}(P(x, L), P(y, L)) \\
&\leq 5d_{\mathscr{H}}(K - x, L - x) + d_{\mathscr{H}}(B(x, 2d_L(x)), B(y, 2d_L(y))) \\
&\leq 5d_{\mathscr{H}}(K, L) + |x - y| + 2|x - y| = 5d_{\mathscr{H}}(K, L) + 3|x - y|,
\end{aligned}$$

and the proof is complete. □

Proof of Lemma 4.3.7 We prove (i).

Suppose (i, a). Fix $(t, x) \in \operatorname{dom} V \cap ((]0, +\infty[\setminus C) \times \Omega)$ and let $(p_t, p_x, q) \in T_{\operatorname{epi} V}(t, x, V(t, x))^-$. From (1.5) and (4.43), we have $(1, f(t, x, \bar{u}), -\ell(t, x, \bar{u})) \in T_{\operatorname{epi} V}(t, x, V(t, x))$. Thus $p_t + \langle p_x, f(t, x, \bar{u}) \rangle - q\ell(t, x, \bar{u}) \leq 0$, and so

$$-p_t + H_{rep}(t, x, -p_x, -q) \geq 0.$$

Suppose next that (i, b) is satisfied and let $j \in \mathbb{N}^+$. By the separation theorem, (i, b) implies that

$$(\{1\} \times \Phi(t, x, \beta)) \cap \text{cl co}\, T_{\text{epi}\, V}(t, x, V(t, x)) \neq \emptyset$$

for all $(t, x) \in \text{dom}\, V \cap ((]0, j[\backslash C) \times \Omega)$ and all $\beta \in \mathbb{R}$. From Lemmata 2.2.5 and 2.2.4, for a set $C_j \subset [0, j]$, with $\mu_{\mathcal{L}}(C_j) = 0$, and for all $t_0 \in [0, j]\backslash C_j$ and all $(x_0, v_0) \in \text{epi}\, V(t_0, .)$ there exists a Φ-trajectory $(x, \beta)(.)$ on $[t_0, j]$, with $(x(t_0), \beta(t_0)) = (x_0, v_0)$, satisfying $(x, \beta)(t) \in \text{epi}\, V(t, .)$ for all $t \in [t_0, j]$ and

$$\emptyset \neq \operatorname*{Lim\,sup}_{\xi \to t_0+} \left\{ \frac{1}{\xi - t_0} (x(\xi) - x_0, \beta(\xi) - \beta(t_0)) \right\} \subset \Phi(t_0, x_0, \beta_0).$$

Taking $\beta_0 = V(t_0, x_0)$, by the Measurable Selection Theorem we conclude that there exist two measurable functions $u(.)$ and $r(.)$, with $u(t) \in U(t)$ and $r(t) \in [0, c(t)(1 + |x(t)|) - \ell(t, x(t), u(t))]$ for a.e. $t \in [t_0, j]$, such that $\beta(t) = V(t_0, x_0) - \int_{t_0}^t \ell(s, x(s), u(s))\, ds - \int_{t_0}^t r(s)\, ds \geq V(t, x(t))$ for any $t \in [t_0, j]$. Then

$$\beta(t) - \beta(t_0) \geq V(t, x(t)) - V(t_0, x_0) \quad \forall t \in [t_0, j].$$

So, dividing by $t - t_0$ the last inequality and passing to the lower limit as $t \to t_0+$, (4.43) follows for $C = \cup_{j \in \mathbb{N}^+} C_j$.

Next we prove (ii).

Assuming (ii, a) and arguing similarly to (i), we can conclude that there exists $C \subset]0, +\infty[$, with $\mu_{\mathcal{L}}(C) = 0$, such that $-p_t + H_{rep}(t, x, -p_x, -q) \leq 0$ for all $(p_t, p_x, q) \in T_{\text{epi}\, V}(t, x, V(t, x))^-$ and all $(t, x) \in \text{dom}\, V \cap ((]0, +\infty[\backslash C) \times \text{int}\, \Omega)$.

Now, assume (ii, b) and let $j \in \mathbb{N}^+$. From Lemmata 4.2.5 and 2.2.5 applied to the set-valued map $\tilde{\Phi}(j - ., ., .)$, and the Measurable Selection Theorem, we can find a subset $C_j \subset [1/j, j]$, with $\mu_{\mathcal{L}}(C_j) = 0$, such that for any $(t_0, x_0) \in (]1/j, j]\backslash C_j) \times \text{int}\, \Omega$ and any $u_0 \in U(t_0)$ there exist $t_1 \in [1/j, t_0[$ and a trajectory-control pair $((x, \beta), (u, r))(.)$ satisfying

$$\begin{cases} (x'(t), \beta'(t)) = (f(t, x(t), u(t)), -\ell(t, x(t), u(t)) - r(t)) & \text{a.e. } t \in [1/j, t_0], \\ (x(t_0), \beta(t_0)) = (x_0, 0), \\ u(t) \in U(t),\ r(t) \in [0, c(t)(1 + |x(t)|) - \ell(t, x(t), u(t))] & \text{a.e. } t \in [1/j, t_0], \\ (x'(t_0), \beta'(t_0)) = (f(t_0, x_0, u_0), -\ell(t_0, x_0, u_0)), \\ x([1/j, t_0]) \subset \text{int}\, \Omega. \end{cases}$$

From Lemma 4.3.11 we get

$$\beta(s) - \beta(t_0) \geq V(s, x(s)) - V(t_0, x(t_0)) \quad \forall s \in [1/j, t_0].$$

Hence, dividing by $t_0 - s$, passing to the lower limit as $s \to t_0-$, and since $u_0 \in U(t_0)$ is arbitrary, we have (4.44) after taking $C = \cup_{j \in \mathbb{N}^+} C_j$. □

A.4 Bibliographic Notes

Infinite horizon control problems have seen advances also in stochastic control and reinforcement learning: discounting and stability were analyzed by Bertsekas et al. [9, 10] and Blackwell [11]. Tube-based methods for handling state constraints were developed by Basco et al. [7, 8], who also studied the Lipschitz regularity of the value function, highlighting the importance of inward-pointing conditions for stability and reliable trajectory planning. Infinite horizon settings appear naturally in economics, e.g., the Ramsey model [22], where discounted utility maximization leads to Pontryagin-type necessary conditions. The co-state variable can be interpreted as a shadow price, linking marginal increases in capital to variations in total optimal utility. Basco et al. [6] derived a normal form of the maximum principle with partial and full sensitivity relations and initial-time transversality conditions. Viscosity solutions form the foundation of HJB theory—see Crandall et al. [14]. Ishii [17, 18] extended this framework to time-measurable Hamiltonians and nonsmooth boundaries. Basco et al. [2–5] proposed a geometric approach based on normals to the epigraph, ensuring uniqueness and applicability to continuous and merely measurable infinite horizon settings.

References

1. Aubin, J.-P., Cellina, A.: Differential Inclusions: Set-valued Maps and Viability Theory, vol. 264. Springer Science & Business Media (2012)
2. Basco, V.: Representation of weak solutions of convex Hamilton-Jacobi-Bellman equations on infinite horizon. J. Optim. Theory Appl. **187**(2), 370–390 (2020)
3. Basco, V.: Weak epigraphical solutions to Hamilton-Jacobi-Bellman equations on infinite horizon. J. Math. Anal. Appl. **515**(2), 126452 (2022)
4. Basco, V.: Control problems on infinite horizon subject to time-dependent pure state constraints. Math. Control Signals Syst. **36**(2), 423–450 (2024)
5. Basco, V.: Locally bounded variations epigraph property of the value function to infinite horizon optimal control problems under state constraints. Bollettino dell'Unione Matematica Italiana **17**(4), 825–846 (2024)
6. Basco, V., Cannarsa, P., Frankowska, H.: Necessary conditions for infinite horizon optimal control problems with state constraints. Math. Control Related Fields **8**, 535–555 (2019)
7. Basco, V., Frankowska, H.: Hamilton-Jacobi-Bellman equations for infinite horizon control problems under state constraints with time-measurable data. NoDEA-Nonlinear Differ. Equ. Appl. **26**(1), 7 (2019)
8. Basco, V., Frankowska, H.: Lipschitz continuity of the value function for the infinite horizon optimal control problem under state constraints. In: Alabau-Boussouira, F., et al. (ed.) Trends in Control Theory and Partial Differential Equations, Volume 32 of Springer INdAM Series, pp. 15–52. Springer International Publishing (2019)

9. Bertsekas, D.: Dynamic Programming and Optimal Control: Volume I, vol. 1. Athena Scientific (2012)
10. Bertsekas, D.: Reinforcement Learning and Optimal Control. Athena Scientific (2019)
11. Blackwell, D.: Discounted dynamic programming. Ann. Math. Stat. **36**(1), 226–235 (1965)
12. Castaing, C., Valadier, M.: Convex Analysis and Measurable Multifunctions. Lecture Notes in Mathematics. Springer, Berlin, Heidelberg (1977)
13. Clarke, F.H.: Optimization and Nonsmooth Analysis, Volume 5 of Classics in Applied Mathematics, 2 edn. Society for Industrial and Applied Mathematics (SIAM), Philadelphia, PA (1990)
14. Crandall, M.G., Evans, L.C., Lions, P.-L.: Some properties of viscosity solutions of Hamilton-Jacobi equations. Trans. Am. Math. Soc. **282**(2), 487–502 (1984)
15. Ekeland, I., Valadier, M.: Representation of set-valued mappings. J. Math. Anal. Appl. **35**(3), 621–629 (1971)
16. Folland, G.B.: Real Analysis: Modern Techniques and Their Applications, vol. 40. Wiley, New York (1999)
17. Ishii, H.: Hamilton-Jacobi equations with discontinuous Hamiltonians on arbitrary open sets. Bull. Fac. Sci. Engrg. Chuo Univ. **28**, 33–77 (1985)
18. Ishii, H., Koike, S.: A new formulation of state constraint problems for first-order PDEs. SIAM J. Control. Optim. **34**(2), 554–571 (1996)
19. Le Donne, A., Marchi, M.V.: Representation of lipschitzian compact-convex valued mappings. Atti della Accademia Nazionale dei Lincei. Classe di Scienze Fisiche, Matematiche e Naturali. Rendiconti **68**(4):278–280 (1980)
20. Lojasiewicz, S.: Parametrizations of convex sets. In: Progress in Approximation Theory (1991)
21. Ornelas, A.: Parametrization of Carathéodory multifunctions. Rendiconti del Seminario Matematico della Università di Padova **83**, 33–44 (1990)
22. Ramsey, F.P.: A mathematical theory of saving. Econ. J. **38**(152), 543–559 (1928)
23. Vinter, R.B.: Optimal Control. Birkhäuser, Boston, MA (2000)

Index

V. Basco, *Nonsmooth Constrained Optimal Control on Infinite Horizon*, SpringerBriefs in Mathematics, https://doi.org/10.1007/978-3-032-08938-0

MIX
Papier aus verantwortungsvollen Quellen
Paper from responsible sources
FSC® C105338

If you have any concerns about our products, you can contact us on
ProductSafety@springernature.com

In case Publisher is established outside the EU, the EU authorized representative is:
Springer Nature Customer Service Center GmbH
Europaplatz 3, 69115 Heidelberg, Germany

Printed by Libri Plureos GmbH
in Hamburg, Germany